A HANDBOOK OF

LABORATORY SOLUTIONS

by M. H. Gabb B Sc
and W. E. Latchem B Sc

Other Related Books

A Text Book of Environmental Science	*Dr. Vidya Thakur*
Basic Ground-Water Hydrology	*U.S.D.I.*
Challenges and Strategies of Dryland Agriculture	*Rao, S.C.*
Handbook of Techniques in Microbiology - A Laboratory Guide to Microbes	*Karwa, A.S.*
Laboratory Manual of Microbiology	*Kumar, V.*
Laboratory Manual of Microbiology, Biochemistry and Molecular Biology	*Saxena, Jyoti*
Laboratory Exercises in Plant Pathology : An Instructional Kit (2 Vol. Set)	*Baudoin, A.B.A.M.*
New Vistas of Organic Farming	*Mukund Joshi*
Organic Agriculture in the Tropics and Subtropics	*Ulrich kopke*
Organic Agriculture	*J.C. Tarafdar et al.*
Organic Farming: Theory & Practice	*SP Palaniappan*
Plant Analysis - Research Methods	*S.S. Narwal*
Plant Analysis: Comprehensive Methods & Protocols	*B.K. Garg*
Principles in the Quantitative Analysis of Waters, Fertilizers, Plants Soils	*U.S. Sree Ramulu*
Resource Conserving Techniques in Crop Production	*A.R. Sharma*
Salinity Tolerance in Plants: Methods, Mechanisms & Management	*B.K. Garg*
Soil and Plant Analysis Laboratory Manual	*John Ryan et al.*
Soil and Plant Analysis	*C.S. Piper*
Soil and Water Conservation in Semi-arid area	*FAO*
Soil Salinity Assessment Methods & Interpretation of Electrical Conductivity Measurements	*FAO*
Soil Survey Laboratory Methods Manual	*USDA*
Soil Survey Manual - Revised and Enlarged Ed.	*USDA*
Soil-Water Engineering Field & Laboratory Manual	*Thomas J. Trout*
Standard Methods for Analysis of Soil Plant & Water	*I.C. Gupta et al.*
The Soil – Plant Systems in Relation to Inorganic Nutrition	*Maurice Fried*
Use of Saline Water in Agriculture	*I.C. Gupta*

A HANDBOOK OF LABORATORY SOLUTIONS

by M. H. Gabb B Sc
and W. E. Latchem B Sc

Edited by Philip Kogan

SCIENTIFIC
PUBLISHERS (INDIA)

Reprint in India, 2013

by

Scientific Publishers (India)

5-A, New Pali Road,
P.O. Box 91
Jodhpur - 342 001 (India)

4806/24, Ansari Road,
Daryaganj
New Delhi - 110 002 (India)

www.scientificpub.com

ISBN: 978-81-7233-806-0

First Published 1967

Printed in India

Contents

Foreword

The purpose of this book is to provide a concise and handy reference guide to the numerous 'recipes' for the making up of chemical solutions used in laboratories. It is intended for the laboratory worker in industry and in research as well as in teaching institutions. It is hoped that it will meet most of the everyday requirements of such workers—in chemistry, physics, biology and in engineering laboratories. The book has been divided into chapters in which preparations of one particular use, or related uses (e.g. histology), are grouped together alphabetically. Where appropriate, the uses of the solution are stated, and cross-reference made. The comprehensive index includes both the names and uses of the solutions covered, and a list of further references is provided.

We should like to acknowledge with thanks the assistance given by Messrs B.D.H. Ltd, George T. Gurr Ltd and May & Baker Ltd, who provided technical literature from which much useful information was drawn.

M. H. Gabb
W. E. Latchem

April 1966

Abbreviations

A = analytical reagent purity
alc = alcoholic
B = ordinary purity
aq = in water
conc = concentrated
D = weight of 1 ml of liquid at 20°C
dil = dilute
Eq = equivalent (weight) of the substance
eth = ethanolic
F.W. = formula weight or molecular weight
g = gram or grams
ml = millilitre
M = molar concentration
mg = milligram
N = normal concentration
ppt = precipitate
R.I. = refractive index
sat = saturated
s.g. = specific gravity
soln = solution
vol = volume
ws = water soluble

The following information, where relevant, is given for the solute used in making each solution: name; alternative name; F.W. and Eq; relationship between them e.g. M = N; D (if liquid); instructions for making the solution; other concentrations in use; specific uses.

Authors' Note

The majority of bench solutions are common to chemistry, biology and other laboratories. To avoid unnecessary duplication, the reader is cross-referred from one section to another, for instructions on the making up of such solutions. The uses are given in each section, however, when specific to a particular discipline. The reader should refer to the index in the first place, because this includes uses as well as names of the solutions.

The modern convention of using 'ethanol' in place of 'alcohol' has been followed except where a particular solution is still commonly known by the alcoholic name. Where ethanol is mentioned and no strength specified, 95% upwards is suitable. Industrial methylated spirit 74° o.p. is a satisfactory substitute in UK. Distilled or de-ionized water is invariably to be used in aqueous solutions unless it is specifically stated to the contrary.

It is necessary to add that where uses of solutions are given, this does not imply that these are the sole uses. Some solutions have so many uses that it would be superfluous to include any.

M. H. G.
W. E. L.

I

Solutions—Basic Definitions

Many of the reagents used in science are in the form of solution, usually in water, but also in other solvents or mixtures of solvents. For many purposes, as in ordinary bench reagents, the exact value of concentration is not of critical importance; in other cases, such as the preparation of a standard solution, the concentration must be as accurate as possible. Solutions of known concentration are called standard solutions. Concentration can be expressed in several ways: as the weight of solute in a given volume of solution; as the weight in a given weight of solvent or solution; by stating the weight of solute as a percentage of the weight of solvent or by stating the density of the solution.

MOLAR AND NORMAL SOLUTIONS

The actual weight of the solute is normally expressed as a fraction of *molar* or *normal*. Weights of solute required for molar or normal solutions are calculated from the *molecular formulae*.

The *symbol* of an element is the letter, or letters, which stand for one atom of it.

The *atomic weight* of an element is the weight of its atom on a scale on which the carbon atom of mass number 12 weighs exactly 12·000 units.*

A list of atomic weights is on page 100. They are usually written by using the symbol of the element, e.g. the atomic weight of carbon, representing the average weight of all the atoms present in it, is written C = 12·011 to three places of decimals, or 12·01115 to five places. Similarly for hydrogen, H = 1·00797 and O = 15·9994, or to three places of decimals, H = 1·008 and O = 15·999.

* The carbon scale replaced the oxygen scale in 1961.

The *formula* of a compound or element shows the composition of its *molecule*, or smallest possible particle.

For example, the formula of glucose, $C_6H_{12}O_6$, shows its molecule to contain six carbon atoms, twelve hydrogen atoms and six oxygen atoms. On the atomic weight scale the molecule will weigh $(6 \times 12{\cdot}011) + (12 \times 1{\cdot}008) + (6 \times 15{\cdot}999) = 180{\cdot}156$ atomic weight units, or 180·159 if the atomic weights used are taken to four places of decimals and the total corrected to three places. The number 180·159 is called the molecular weight or formula weight of glucose.

Although it consists of a conglomeration of ions, the conventional formula of sodium chloride is written NaCl. On the atomic weight scale this weighs 22·9898 (Na) + 35·453 (Cl) = 58·443. As, strictly speaking, a molecule of sodium chloride does not exist, 58·443 should always be called the *formula weight.* For the sake of uniformity the term formula weight, defined below, is used throughout for all substances.

Formula weight of a substance is the sum of the atomic weights of the atoms present in its conventional formula.

Formula weights such as 58·443 for sodium chloride are all numbers; 58·443 grams of sodium chloride is called its *gram formula weight* or ONE MOLE.

A *molar solution* is one which contains one mole of the dissolved substance in a litre of solution.

The concentration of any solution can be expressed as a fraction of molar and this fraction is called the molarity of the solution. For example, a solution of glucose containing 18·0159 g per litre is said to be 0·1 M, one-tenth molar, decimolar or $\frac{M}{10}$.

A *normal solution* is one which contains the gram-equivalent of the dissolved substance in a litre of solution.

The *gram-equivalent* is the weight in grams which combines with or displaces 8 (strictly 7·9997) g of oxygen, or the gram-equivalent of any other substance, for example 1·00797 g of hydrogen.

Concentrations are expressed as *molarities.* For those who wish to use normalities (concentrations expressed as fractions of normal) the relationship between the two is shown in the text as M = N etc. For example, sulphuric acid, H_2SO_4, has a formula weight of 98·078. A solution containing 98·078 g per litre is molar. Its equivalent, the number of grams which contain 1·008 g of replaceable or ionizable hydrogen, is 49·039; a solution containing

98·078 g per litre is therefore 2 N. This is shown in the text by M = 2 N.

Except where concentration is quoted as a percentage, the units of concentration used are indicated; a percentage concentration means that this number of grams is present in 100 ml of the solvent. For example, a 5% solution in ethanol means that 5 g of solute are dissolved in 100 ml of ethanol; unless otherwise stated, the ethanol intended is industrial spirit. The water used for making solutions is distilled, or de-ionized, unless otherwise stated.

STANDARD SOLUTIONS

A *standard solution* is one of known concentration, expressed as a fraction of molar or normal.

Substances which are pure enough to be weighed out and dissolved in distilled or de-ionized water to give a solution of accurately known concentration are called *primary standards*. For convenience in weighing they should, for preference, be solids and should have the following characteristics:

they must be obtainable in a very pure state;
readily soluble in the solvent to be used;
stable and unchanged in air at ordinary or moderate temperatures;
of reasonably high formula weight and, preferably, colourless in solution.

A list of primary standards is given on page 12.

Slightly less suitable substances, sometimes called secondary standards, are also listed on page 14.

To make a standard solution a suitable primary standard is chosen and its formula weight is taken from the list. The correct weight required to make a solution of the concentration and volume needed is calculated and this amount is carefully and accurately weighed. For example, the formula weight of sodium chloride is 58·443; a litre of molar solution would need 58·443 g of the pure solid. A litre of 0·1 M solution needs 5·8443 g and if only 250 ml of 0·1 M solution is to be made, the amount weighed would be a quarter of 5·8443, or 1·4611 g. How accurately this is weighed depends on how accurate the concentration of the solution needs to be.

THE PURITY OF CHEMICAL SUBSTANCES

Chemicals are manufactured to three grades of purity. Many are available in a very pure state suitable for use in analysis. A specification on the label indicates the maximum limits of all the possible impurities and the minimum percentage of the substance itself, obtained by analysing samples of it after manufacture. These are known by trade names such as AnalaR and Pronalys.

Nearly all other chemical substances can be obtained in a state only slightly less pure, with the same type of specification on the label. The third grade is 'Technical', of a much lower degree of purity and usually with no specification. The letters 'B.P.' following the name indicates that the substance complies with the British Pharmacopœia standard of purity.

2

Solutions—Handling Techniques

CLEAN APPARATUS

The preparation of solutions of known concentration requires accurately graduated glass apparatus which is chemically clean and free from grease. Precise measurement of volume is impossible unless the graduations are accurate; the presence of other substances on the surface of dirty glassware will interfere with the purity of the solution being made; grease distorts the meniscus, or curved surface, of the solution and makes it difficult to read the volume accurately; drops of solution are left behind on the inside surface of the apparatus if it is greasy, causing errors in the volume.

Directions for making cleaning solutions are given on page 11. If the apparatus is in regular use washing it with the detergent solution may be enough. If, after this, the surface is still greasy, with drops of water clinging to it, the dichromate solution should be tried, wetting the whole of the inside surface with it or leaving the apparatus full of it overnight.

If this is unsuccessful, sodium hydroxide solution, potassium hydroxide in ethanol and potassium permanganate are tried in that order. In each case the apparatus is washed thoroughly with water and then distilled water afterwards. If permanganate is used, the brown stains of manganese dioxide are removed with concentrated hydrochloric acid before washing with water.

Non-measuring apparatus made of thin glass can be made chemically clean and free from grease by passing steam through it for some time. Thick glass may crack and graduated apparatus should never be filled with anything hot because expansion will affect the volume for a considerable time. Clean apparatus can be dried by a stream of hot air and should be protected from contamination by dust by stoppering or covering it. Traces of water can be removed before using the stream of air by rinsing with a little acetone.

MEASURING OR GRADUATED APPARATUS

The apparatus commonly used in making solutions consists of graduated flasks, burettes and pipettes. Measuring cylinders have a limited use in making solutions of less precise concentration. There are three grades of accuracy each of which allows a maximum possible error, or tolerance, in the piece of apparatus. The tolerance is least for A standard of accuracy, greater for B standard and of any size for ungraded glassware. Maximum tolerances for flasks, burettes and pipettes are given in Appendix 1.

Apparatus is labelled to indicate the standard of accuracy and other essential details. On older apparatus, for example, '250 ml C 20°C B' means 'Contains 250 ml at 20°C with B standard of accuracy'; '25 ml D 20°C A' means 'Delivers 25 ml of liquid at 20°C with A standard of accuracy'. More recent graduated glassware has the letters C and D replaced by IN for 'contains' and EX for 'delivers'.

THE PIPETTE

A pipette is a glass tube with a bulb half way along it and with one end drawn off to a narrow tip. The graduation mark is between the wider end and the bulb.

If free from grease and chemically clean it is washed out with distilled water, drained by allowing the water to run out of the tip, wiped dry on the outside and the last drop of water is blown out.

The tip is placed in the solution and 2 or 3 ml of it are drawn into the pipette by gently sucking at the wide end. The solution is held in the pipette by placing the forefinger over the wide end. By holding the pipette horizontally the solution is run gently over the whole of the inside surface beyond the graduation mark but not so as to wet the wide end at which the mouth is applied. The solution is allowed to drain out of the pipette and the process is repeated.

The solution is now drawn into the pipette until the meniscus is higher than the graduation mark, the forefinger is applied to hold the solution in and pressed firmly on the wide end. With the pipette vertical and the eye directly opposite the meniscus, the pressure is released without actually taking the forefinger off and renewed when the meniscus has fallen so that it just rests on, or touches, the graduation mark.

The tip of the pipette is touched against the inside of a beaker to remove any part of a drop suspended from it and the solution is then allowed to drain out into the vessel in which it is required. If the pipette is labelled 'D' it should be held with the tip touching the inside surface of the vessel for fifteen seconds after the last drop has drained out in the normal way. This was the method used when it was graduated. Those labelled 'EX' should be allowed to drain for two or three seconds after the solution stops running out.

THE BURETTE

The burette is a graduated glass tube from which liquids may be delivered in accurate quantities by means of a glass tap at the bottom. Burettes are graduated in millilitres and tenths of millilitres starting with the zero mark near the top. The tap at the bottom leads to a narrow jet or tip. The glass stopcock is cleaned, dried, coated with a thin layer of petroleum jelly and inserted into the dry burette. Rotating the stopcock should make the tap almost transparent without using so much petroleum jelly as to block the stopcock.

Petroleum jelly allows the stopcock to turn freely but occasionally some of it is carried by the flow of liquid into the tip causing it to become blocked up. The tip should be held in hot water to melt the obstruction which is then washed out of the tip by opening the tap to allow solution or water to run through.

If this is unsuccessful, all the liquid in the tip can be drawn out by attaching a filter pump to it. If this is done during use, the tip is exactly refilled with solution by carefully opening the tap. The burette reading is taken before and afterwards and the difference is subtracted from the final reading.

The stopcocks of burettes sometimes become stuck, especially if they are not washed thoroughly after using alkali in them. Running hot water over the outside of the tap will sometimes make it expand enough to free it, because the stopcock remains cold and of the same size. The same principle can be applied more riskily by using a small flame to heat the outside of the tap. Another method is to immerse the tap in warm water for some time or to use penetrating oil. Plastic diaphragm taps avoid sticking and do not need grease.

The burette is washed with distilled water before use and drained.

About 5 ml of the solution to be used in it is put in and the whole of the inside surface is wetted with it before allowing it to drain out through the tap. This washing and draining process is repeated with another 5 ml and the burette is filled with the solution to a point about 1 cm above the zero mark. The tap is opened to fill the tip and to bring the curved meniscus of the solution on to the graduated scale. The burette is then supported in a vertical position in a stand, checking its vertical position by lining it up with two vertical corners of the room.

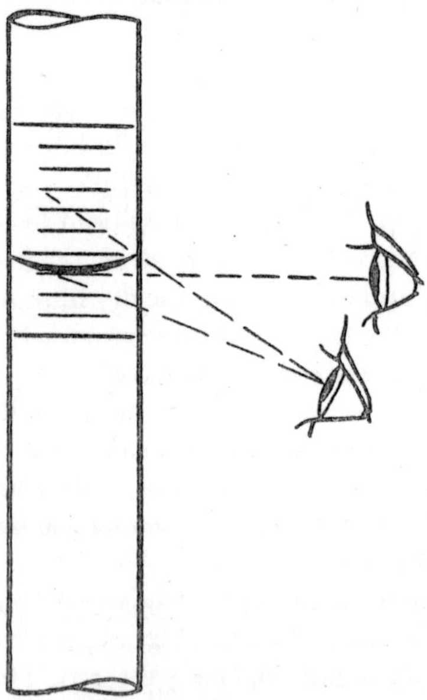

FIG. 1

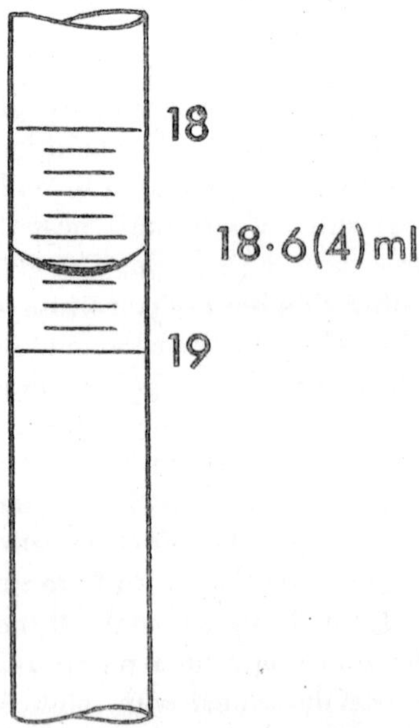

FIG. 2

The stopcock must always be pressed slightly inwards when it is turned. This is best done by putting the left hand round the stem and turning the stopcock with the left thumb and forefinger. This automatically pulls the stopcock into the main body of the burette and also leaves the right hand free for swirling the solution into which the liquid is running.

The bottom of the meniscus is used to take a reading with the eye exactly level with it to avoid errors caused by parallax (Fig. 1). It is worth while recording an estimated second place of decimals, including it in brackets, e.g. 18·6(4) ml in Fig. 2. It is preferable to use the full volume in the burette instead of refilling it to the zero mark after using half of the volume.

MAKING A SOLUTION OF APPROXIMATE CONCENTRATION

Slightly more than the weight indicated by the formula of the dissolved substance is weighed approximately. If the dissolved substance is a liquid it can be measured accurately enough by a measuring cylinder. It is dissolved in distilled water in a beaker and made up to the volume required. Concentrated sulphuric acid, sodium hydroxide and several other substances produce heat on solution. It is preferable to dissolve them with stirring and to cool the solution before making it up to the correct volume. It is advisable to get into the habit of pouring *all* concentrated acids into water, rather than the reverse, although only concentrated sulphuric acid produces so much heat that pouring water into it is dangerous. Always shake a solution before use to make the concentration the same throughout.

MAKING STANDARD SOLUTIONS BY WEIGHING

The pure solid is weighed on a watch glass or in a weighing bottle or a scoop. If the weight is small the solid is washed straight into the graduated flask through a clean funnel, using a jet of distilled water from a wash bottle. If the weight is large, or if the substance must be heated in the water to make it dissolve rapidly, it is washed into a beaker and made into a solution. This is cooled if necessary and poured into the flask.

This is often done by holding a glass rod in the mouth of the flask as shown in Fig 3; as the solution is poured in, the lip of the beaker is held so that it just touches the glass rod. All the solid which was weighed must find its way into the final solution; the vessel in which it is weighed must be carefully washed out into the beaker or flask and so must the beaker. Water is added to the solution in the flask until the curved meniscus of the liquid just touches, or rests on, the graduation mark. To avoid overshooting the mark the last few drops are added with a dropper or teat pipette.

The solution is shaken vigorously to equalize the concentration throughout. If the solution is to be used from a beaker or any other vessel this must be washed out with a few millilitres of the solution before filling it, so that moisture on the inside surface will not alter the concentration.

A solution must always be labelled with the name of the dissolved substance, its concentration and the date.

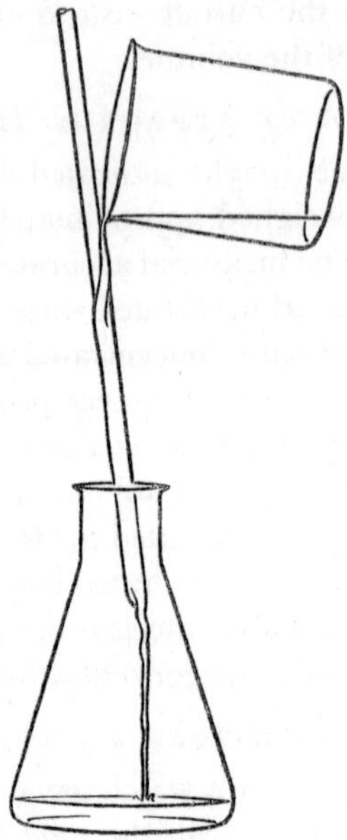

FIG. 3

STANDARDIZATION OF SOLUTIONS BY TITRATION

Because the number of primary standards is small, most solutions are made up to approximately the correct concentration and their exact concentration determined by titration with a standard solution of one of the primary standards, made by weighing (see above).

The method of carrying out a titration is as follows. A clean, grease-free burette is washed with distilled water and allowed to drain through the tap. About 5 ml of the solution to be used in it is added and the whole of the inside surface is wetted with it before letting it drain out through the tap. The process is repeated with another 4 or 5 ml of the solution, and after draining the burette is filled with the solution above the zero mark. The tap is opened to fill the tip and to bring the meniscus of the solution on to the graduated part of the burette. A funnel is not really necessary in filling the burette but if one is used it must be removed before any readings are taken.

The pipette is washed and filled with solution exactly as previously described and the solution is transferred to a conical flask, previously washed with distilled water. The standard solution may be either in the burette or the pipette. A suitable quantity of indicator is added to the solution in the flask and, after taking the reading on the burette, the solution is run from it into the flask.

When the indicator shows signs of changing colour further additions are made in drops until the colour change is complete. After taking the reading the whole procedure, which is called a *titration*, is repeated until agreeing results, within 0·1ml of each other, have been obtained.

A drop is more than enough to complete the colour change in some cases, but a fraction of a drop can be formed on the tip of the burette and added to the flask by allowing the tip to touch its inside surface. The inside surface can be washed with a jet of water from a wash bottle to ensure that all the splashes which form on it during titration enter the main body of the solution. The concentration of the unknown solution is calculated.

CLEANING SOLUTIONS

Detergent solution. Shake approx 20 g of solid detergent in a litre of water and add a little conc nitric acid. Use 20 ml of liquid detergent to make the same volume.

Sodium dichromate in sulphuric acid. Dissolve 10 g of sodium dichromate in approx 15 ml of water and make the solution up to 100 ml by adding, slowly and with cooling, concentrated sulphuric acid. Treat with the same care as the conc acid.

Potassium permanganate. Dissolve by heating 10 g of solid and make up to 1 litre. This solution can also be made alkaline with M sodium carbonate solution before use. Brown stains are removed from the apparatus after use with conc hydrochloric acid.

Sodium hydroxide. Use 2 M solution (see chapter 4)

Potassium hydroxide in ethanol. Dissolve 56 g of solid in 1 litre of industrial spirit.

Solutions for the removal of stains

Carbon deposits: 6 g of trisodium phosphate and 3 g of sodium oleate in 100 ml of water.

Iodine: sodium thiosulphate solution.

Indelible pencil: acetone.

Iron stains: dilute or conc hydrochloric acid.

Sulphur: ammonium sulphide solution.

3

Solutions for Titrations

PRIMARY STANDARDS

Standard solutions of the following substances can be made by direct weighing. Only reagents of analytical standard of purity should be used.

Potassium dichromate. $K_2Cr_2O_7$. F.W.=294·19 Eq=49·03 M = 6 N. For direct titration and for standardization of sodium thiosulphate.

Potassium iodate. KIO_3. F.W.=214·01 Eq = 35·67 M = 6N. For direct titration and standardization of sodium thiosulphate.

Sodium carbonate (anhydrous). Na_2CO_3. F.W. = 105·99 Eq = 53·00 M = 2 N. For standardization of solutions of strong acids.

Sodium chloride. NaCl. F.W.=58·44 Eq=58·44 M=N. For standardization of silver nitrate solution.

Sodium oxalate. $(COONa)_2$. F.W.=134·00 Eq=67·00 M= 2 N. For standardization of potassium permanganate and ceric sulphate solutions.

These are all in the form of anhydrous solids which may be dried at temperatures between 120°C and 180°C without decomposition. The correct weight for the solution required is calculated from the value of F.W. or Eq. This weight of the dry substance is accurately weighed, dissolved in distilled water and made up to the correct volume. (Preparation of a standard solution by weighing, page 9.)

The substances listed below are also used as primary standards. They may gain or lose water from the air but are acceptable for all work except that of the highest accuracy, provided they are carefully stored.

1. For the standardization of solutions of strong acids

Guanidine carbonate. $[(NH_2)_2.C.NH]_2.H_2CO_3$. F.W. = 180·17 Eq = 90·09 M = 2 N.

Sodium tetraborate, borax. F.W. = 381·37 Eq = 190·68 M = 2 N.

2. For the standardization of alkaline solutions

Benzoic acid. C_6H_5COOH. F.W. = 122·13 Eq = 122·13 M = N.

Oxalic acid. $(COOH)_2.2H_2O$. F.W. = 126·07 Eq = 63·04 M = 2 N.

Potassium hydrogen phthalate. $COOH.C_6H_4.COOK$. F.W. = 204·23 Eq = 204·23 M = N.

Potassium hydrogen tartrate. COOH.CH(OH).COOK. F.W. = 158·16 Eq = 158·16 M = N.

Sulphamic acid. NH_2SO_3H. F.W. = 97·09 Eq = 97·09 M = N.

3. For the standardization of oxidizing agents

Ammonium ferrous sulphate. $(NH_4)_2SO_4.FeSO_4.6H_2O$. F.W. = 392·14 Eq = 392·14 M = N. Should be made up in dil sulphuric acid.

Arsenous oxide. As_2O_3. F.W. = 197·84 Eq = 49·46 M = 4 N. Made up in dil sodium hydroxide and the solution is then neutralized or made slightly acid with dil hydrochloric acid.

Hydrazine sulphate. $NH_2.NH_2.H_2SO_4$. F.W. = 130·12 Eq = 32·53 M = 4 N.

Oxalic acid. See standardization of alkaline solutions.

4. For precipitation reactions

Silver nitrate. $AgNO_3$. F.W. = 169·88 Eq = 169·88 M = N.

5. *For iodine titrations*

Antimony potassium tartrate. $KSbO.C_4H_4O_6$. F.W. = 324·92 Eq = 162·46.

These may all be weighed directly. An M solution of sodium chloride will require 58·43 g of the dry solid per litre of solution. To make 500 ml of 0·1 M or 0·1 N solution, 2·922 g will be weighed; unless an exactly 0·1 M solution is required it is easier to weigh about 2·9 g of the salt accurately, make it up to 500 ml of solution and calculate the normality or molarity of the solution.

The following solutions cannot be made up accurately by direct weighing; details are given for making up a solution of approximately molar concentration and the relationship between molar and normal is indicated by M = x N. For example, M = 2 N indicates that a molar solution is twice normal, or 2 N in concentration. Brief details of the method of standardization are given.

ACIDS AND ALKALIS

Acetic acid. CH_3COOH. F.W. = 60·05 Eq = 60·05 M = N D = 1·048 A = at least 99·6% acid. 1 litre of the pure glacial acid weighs 1048 g and is, therefore $\frac{1048}{60 \cdot 05}$ M = 17·3 M = 17·3 N. Dissolve 58 ml of glacial acid in aq and make up to 1 litre. Use standard sodium hydroxide soln with phenolphthalein indicator.

Boric acid. H_3BO_3. F.W. = 61·83 Eq = 61·83 M = N. 61·83 g dissolved in aq and made up to 1 litre. Use boric acid soln with mannitol or glycerol with standard hydroxide soln using sodium phenolphthalein.

Hydrochloric acid. HCl. F.W. = 36·46 Eq = 36·46 M = N D = 1·18 A (conc acid) = 36 % acid approx Molarity = 11·7. 86 ml of conc acid into 500–600 ml of water make up to 1 litre. Use standard sodium carbonate soln with methyl orange or sodium hydroxide with methyl red.

From constant boiling acid. Distil rapidly 1 litre of conc acid neglecting the first 700 ml and the last 50 ml. Measure barometric pressure to the nearest millimetre and read off the concentration of the constant boiling mixture from the standard table below.*

Barometric pressure mm	% HCl by weight	Weight of soln giving one mole of HCl (in g)
770	20·197	180·407
760	20·221	180·193
750	20·245	179·979
740	20·269	179·766
730	20·293	179·555

Nitric acid. HNO_3. F.W. = 63·01 Eq = 63·01 M = N A (conc acid) = 70 % approx D = 1·41 Molarity = 15·8. Pour 64 ml

* Hollingsworth and Foulk.

into water and make the soln up to 1 litre. Standardization as for hydrochloric acid.

Phosphoric acid. H_3PO_4. F.W. = 98·00 Eq = 32·67 M = 3 N A = at least 88 % acid D = 1·75 Molarity = 15·7. Pour 66 ml of syrupy acid into 500 ml of water, stir vigorously and make up to 1 litre. Standard sodium hydroxide soln with methyl orange.

Sulphuric acid. H_2SO_4. F.W. = 98·08 Eq = 49·04 M = 2 N A = at least 97% acid D = 1·84 Molarity = 18·5. Pour 55 ml of conc acid slowly into 500 ml of water and make up to 1 litre. Standardization as for hydrochloric acid.

Ammonia solution (ammonium hydroxide). F.W. = 17·03 Eq = 17·03 M = N A = at least 30–35% ammonia D = 0·88 Molarity = 18. Pour 56 ml into 500 ml approx of water and make up to 1 litre. Standard hydrochloric acid with methyl red indicator.

Barium hydroxide (baryta). $Ba(OH)_2.8H_2O$. F.W. = 315·47 Eq = 157·74 M = 2 N. Not sufficiently soluble to make a molar or normal solution possible. Readily absorbs carbon dioxide from the air and is not much used. Solubility 3·89 g per 100 g at 20° C.

Potassium hydroxide. KOH. F.W. = 56·11 Eq = 56·11 M = N. Weigh rapidly in a closed vessel about 56·5 g (see sodium hydroxide), dissolve in freshly boiled and cooled distilled water with stirring and make up to 1 litre. Keep the solution in a bottle fitted with rubber stopper or cork. Standard hydrochloric or sulphuric acid with methyl red indicator.

Sodium hydroxide. NaOH. F.W. = 40·00 Eq = 40·00 M = N. As for potassium hydroxide, using 41 g of the solid. In both cases the presence of carbonate in the hydroxide solution is decreased by weighing approx 50 g of sodium hydroxide or 65 g of potassium hydroxide. The carbonate is then rapidly washed off the surface of the hydroxide with water before solution.

Carbonate-free sodium hydroxide solution. (*a*) Dissolve the weighed hydroxide in a little ethanol in which the carbonate does not dissolve, filter and make the filtrate up to the correct volume with water. (*b*) Dissolve the hydroxide in water and add, in drops, barium hydroxide soln to ppt barium carbonate, filter and dilute the soln.

The taps of burettes in which either solution is used must be thoroughly washed after use because the solutions attack glass.

Potassium carbonate. K_2CO_3. F.W. = 138·21 Eq = 69·11 M = 2 N (see sodium bicarbonate).

Potassium bicarbonate (potassium hydrogen carbonate). F.W. = 100·12 Eq = 100·12 M = N (see sodium bicarbonate).

Sodium bicarbonate (sodium hydrogen carbonate). F.W. = 84·01 Eq = 84·01 M = N. All three of these may be weighed directly to give solutions of reasonably accurate concentration.

SOLUTIONS FOR REDOX REACTIONS

Quantities are given for one litre of 0·1 M soln and the relationship M = *x* N can be used to work out the weight needed for 1 litre of 0·1 normal soln.

1. Oxidizing agents or oxidants

Ceric sulphate. $Ce(SO_4)_2.4H_2O$. F.W. = 404·3; anhydrous = 332·24 M=N. Ceric compounds are readily available in the technical grade of purity and are expensive in higher grades of purity. Several methods are suggested for making up solutions from technical grade compounds and two are given.

1. Add 50 g of the oxide in small quantities at a time to 200 ml of a mixture of equal volumes of concentrated sulphuric acid and water, stirring the mixture after each addition. Keep the temperature below 100°C and after stirring continuously for about ten minutes at 100°, cool the solution, dilute to 800 ml with water, filter. Make the solution up to 1 litre with 1 M sulphuric acid. Standardize using arsenous oxide solution or sodium oxalate solution using ferroin indicator.
2. Dissolve approx. 56 g of technical grade sulphate in about 800 ml of 2 N sulphuric acid, boil to obtain a clear, or nearly clear, solution, cool and filter. Dilute with aq to 1 litre. Standardize as above.

Iodine. I_2. F.W. = 253·81 Eq = 126·91 M = 2 N. Add 12·7 g of iodine crystals to a solution of 40 g of potassium iodide in approx. 300 ml of water and shake until all the iodine has dissolved. Make up to 1 litre. Standardize by titration with arsenous oxide solution diluted with an equal volume of water using starch as indicator.

Potassium bromate. See primary standards.

Potassium dichromate. See primary standards.

Potassium iodate. See primary standards.

Potassium permanganate. $KMnO_4$. F.W. = 158·04 Eq = 31·61 M = 5 N. Add approx 3·2 g of crystals, analytical purity, to 300 ml approx of water, heat and stir to dissolve the solid, cool and make up to 1 litre. Allow the soln to stand for 24 hours, filter through an asbestos pad. Standardize with sodium oxalate soln above 70°C. Self-indicating.

Sodium hypochlorite. NaClO. F.W. = 52·47 Eq = 26·24 M = 2 N. Commercial solutions are sold of 10–14 % concentration. Dilute 220 ml of such a solution to 1 litre. Stable for one week in the dark. Standardize with arsenous oxide soln using Bordeaux indicator.

2. Reducing agents or reductants

Ammonium ferrous sulphate. F.W. = 392·14 Eq = 392·16 M = N. Dissolve 39·216 g of crystals in approx 200 ml of 1 M sulphuric acid and make up to 1 litre. Sulphuric acid is necessary to reduce atmospheric oxidation. See primary standards.

Arsenous oxide (sodium arsenite soln). See primary standards. Dissolve 4·945 g in 20 ml of 2 M sodium hydroxide soln and add 2 M hydrochloric or M sulphuric acid until the soln is just acid. Make up to 1 litre.

Mercurous nitrate. $Hg_2(NO_3)_2.2H_2O$ or $HgNO_3.H_2O$ F.W. (second formula) = 280·61 Eq = 280·61 M = N. Dissolve 28·061 g in a litre of 0·5 M nitric acid, or use slightly more solid and standardize using ferric alum soln with potassium thiocyanate soln as indicator.

Sodium thiosulphate. $Na_2S_2O_3.5H_2O$. F.W. = 248·18 Eq = 248·18 M = N. Dissolve 25 g in freshly boiled, cold distilled water and make up to 1 litre. Addition of 0·1 g of sodium bicarbonate is advocated in some texts. Standardize with 25 ml of standard potassium dichromate soln to which has been added 10 ml of 10% potassium iodide soln and 10 ml of M sulphuric acid, using starch as indicator.

Titanous chloride. $TiCl_3$. F.W. = 154·26 Eq = 154·26 M = N. Boil 100 ml of the commercial soln with 100 ml of conc hydrochloric acid, cool and make up to 1 litre.

PRECIPITATION TITRATIONS

Potassium thiocyanate. KCNS. F.W. = 97·18 Eq = 97·18 M = N. Dissolve 10 g of crystals in water and make up to 1 litre.

Standardize with silver nitrate solution acidified with 20 ml of 2 N nitric acid per 25 ml, using ferric alum indicator.

Ammonium thiocyanate. As potassium thiocyanate, using 8 g of crystals.

Sodium chloride. See primary standards. Use 5·845 g per litre.

Silver nitrate. See primary standards. Dry A grade crystals at 110°C for 2 hours. Use 16·988 g per litre. It is cheaper to use 17 g of B grade crystals per litre and standardize the soln with sodium chloride soln, using potassium chromate indicator or an adsorption indicator.

MISCELLANEOUS TITRATION SOLUTIONS

EDTA solution (diamino-ethane-tetra-acetic acid disodium salt). $[CH_2N(CH_2.COOH).CH_2.COONa]_2.2H_2O$. F.W. = 372·25 Eq = 372·25 M = N. For 0·1 M soln dissolve 37·3 g in 1 litre of water. Standardize with standard calcium ion soln at pH 10 using special buffer soln and Solochrome Black T indicator in the presence of a known concentration of magnesium ion.

Potassium ferrocyanide. $K_4Fe(CN)_6.3H_2O$. F.W. = 422·41 Eq = 422·41 is equivalent to 98·07 of zinc. Dissolve 42·241 g in water add about 1 g of sodium carbonate and dilute to 1 litre.

Soap solution. Dissolve 10 g of Castile soap in 100 ml of ethanol by warming it in a beaker of hot water. Add another 500 ml of ethanol and dilute to 1 litre. Filter after standing for a day or two. Standardize with standard calcium ion soln.

Calcium ion solution. To 1 g of pure dry calcium carbonate add a little water and enough 2 N hydrochloric acid in drops to dissolve the chalk, boil, cool and make up to 1 litre. 1 ml = 0·004 g of calcium ion.

Zimmermann-Reinhardt solution. This is used for the titration of ferrous ion in the presence of hydrochloric acid using permanganate solution. Dissolve 70 g of hydrated manganous sulphate crystals in approx 500 ml of water, add in succession, with stirring, 125 ml of conc sulphuric acid and 125 ml of syrupy phosphoric acid (85%) and dilute to 1 litre. Use 25 ml per titration with approx 0·1 N soln.

Clark's soap solution. Dissolve 100 g of Castile soap in 800 ml of ethanol and 200 ml of water, leave it to stand for 24 hours, decant off the liquid and titrate against standard calcium solution. Dilute with the same 80% ethanol until 1 ml = 1 mg of chalk, including the lather factor, which is the volume of soap solution needed to make a permanent lather with 50 ml of distilled water.

Boutron-Boudet soap solution. Dissolve 100 g of Castile soap in 2500 ml of 56% ethanol (1400 ml ethanol + 1100 ml water), titrate against 40 ml of a solution of 0·59 g of barium nitrate per litre. Dilute so that 2·4 ml of the soap solution after dilution gives a permanent lather with 40 ml of the barium nitrate soln. 2·4 ml = 220 parts per million calcium carbonate = 22 French degrees of hardness.

4

Bench Solutions

These need not be made up to the same standard of accuracy as those for titrimetric analysis. Quantities are given in most cases for 1 litre of the most commonly used concentration. Winchester Quart bottles hold 2½ litres and a permanent 2½ litre mark can be put on the bottle by using a measuring cylinder, so that greater volumes can be made up at a time.

The quantities needed for 2½ litres or for other concentrations can be calculated from those given in the text.

ACIDS

Acetic acid glacial. See solutions for titration.

Acetic acid dil. M = N. Dissolve 58 ml of glacial acetic acid in water. Make up to 1 litre.

Acetic acid conc. 5 M = 5 N. A 5 M solution (290 ml of glacial acid per litre) is sometimes used as 'concentrated acid'.

Aqua regia. Add 1 volume of conc nitric acid to 3 volumes of conc hydochloric acid. It should be made up as required.

Hydrochloric acid dil. 2 M = 2 N. Pour 172 ml of conc acid into water. Make up to 1 litre. Use 344 ml of conc acid for 4 M = 4 N.

Nitric acid dil. 2 M = 2 N. Pour 128 ml of conc acid into water. Make up to 1 litre. Use 256 ml of conc acid for 4 M = 4 N.

Orthophosphoric acid dil. M = 3 N. Pour 66 ml of syrupy conc acid into water. Make up to 1 litre.

Sulphuric acid dil. M = 2 N. Pour 55 ml of conc acid *slowly* into about 800 ml of water. Make up to 1 litre. Use 110 ml for 2 M = 4 N. (N.B. Always add the acid to the water.)

ALKALIS

Ammonia solution (ammonium hydroxide). 2 M = 2 N. Pour 112 ml of 0·88 ammonia soln into water. Make up to 1 litre. Use

224 ml for 4 M = 4 N. Avoid getting ammonia into the eyes or breathing it when making the soln.

Barium hydroxide (baryta water). 0·2 M = 0·4 N. Add approx. 70 g of the hydrated hydroxide to 1 litre of water, shake vigorously to form a sat soln. Allow to settle. Protect from atmospheric carbon dioxide by a soda lime tube at any air inlet.

Calcium hydroxide (lime water). Solubility 1·5 g per litre. 0·02 M = 0·04 N. Shake 25 g of hydroxide with 1 litre of water, allow the solid to settle. Protect from atmospheric carbon dioxide as for barium hydroxide. Add more water as soln is used and shake it. Allow to settle before use.

Lithium hydroxide. Approx. M = N. Dissolve 20 g of the hydroxide and 50 g potassium nitrate in water. Make up to 1 litre.

Potassium hydroxide. 2 M = 2 N. Weigh approx 58 g of solid, preferably pellets, dissolve in approx 800 ml of water, with constant stirring. Make up to 1 litre. Keep in a rubber-stoppered or plastic-stoppered bottle. Heat produced by dissolving the solid can be enough to crack a thick glass vessel. 4 M = 4 N is also used.

Sodium hydroxide. 2 M = 2 N. Dissolve 41 g of solid, preferably pellets, as for potassium hydroxide.

Alcoholic potash (potassium hydroxide in ethanol). Dissolve 56 g of pellets or sticks of the solid in ethanol and make up to 1 litre.

Potassium hydroxide in methanol. As for potassium hydroxide in ethanol.

OTHER INORGANIC REAGENTS

Ammonium acetate. CH_3COONH_4. F.W. = 77·08 3 M = 3 N. Dissolve 231 g of dry crystals in water. Make up to 1 litre.

Ammonium carbonate (this is a mixture of ammonium bicarbonate, NH_4HCO_3 and ammonium carbamate, NH_4COONH_2). F.W. approx 79·2 M = 2 N. Dissolve 160 g of solid in approx 500 ml of water, add 120 ml of 0·88 ammonia soln and make up to 1 litre.

Ammonium chloride. NH_4Cl. F.W. = 53·49 3 M = 3 N. Dissolve 160 g in water. Make up to 1 litre. For 5 M use 231 g per litre.

Ammonium molybdate. $(NH_4)_6Mo_7O_{24}.4H_2O$. F.W. = 1235·9. 0·5 M.

Method 1. It is made up in alkaline soln and the necessary nitric acid added. The soln is relatively unstable and forms a ppt on the inside of the bottle. Dissolve 75 g of the salt in 50 ml of 0·88 ammonia soln and 25 ml of water. Dilute to 500 ml and pour slowly and with stirring into 500 ml of 5 M nitric acid.

Method 2. Make up the first solution as above and use equal volumes of it and of 5 M nitric acid as required, keeping the two solutions in separate bottles.

Ammonium nitrate. NH_4NO_3. F.W.=80·05 M=N. Dissolve 80 g of crystals in water. Make up to 1 litre.

Ammonium mercuri-thiocyanate. $(NH_4)_2Hg(SCN)_4$. Approx 0·3 M. Dissolve 9 g of ammonium thiocyanate in approx 50 ml of water, add 8 g of mercuric chloride, stir until dissolved. Make up to 100 ml.

Ammonium oxalate. $(COONH_4)_2.H_2O$. F.W.=142·11 0·25 M = 0·5 N. Dissolve 34 g of crystals in water. Make up to 1 litre.

Ammonium sulphate. $(NH_4)_2SO_4$. F.W.=132·14 M=2 N. Dissolve 132 g in water. Make up to 1 litre.

Ammonium sulphide. $(NH_4)_2S$. F.W. = 68. Ammonium sulphide $(NH_4)_2S$ can be bought as a colourless liquid or a yellow soln containing dissolved sulphur. It is more convenient to buy the soln than to make it. It is sold as 7–10% by weight of H_2S. 10% = approx 3 M = 6 N.

(a) Colourless solution. Cool 200 ml of 0·88 ammonia soln by standing it in ice. Saturate it with H_2S, add 200 ml of 0·88 ammonia soln. Make up to 1 litre.

(b) Yellow solution. Dissolve 10 g of flowers of sulphur in solution obtained above, preferably after the first saturation with hydrogen sulphide.

Ammonium thiocyanate. NH_4SCN. F.W. = 76·12 0·5 M = 0·5 N. Dissolve 38 g of salt and make up to 1 litre with water.

Barium chloride. $BaCl_2.2H_2O$. F.W. = 244·28 0·25 M = 0·5 N. Dissolve 61 g and make up to 1 litre.

Barium nitrate. $Ba(NO_3)_2$. F.W. = 261·35 0·25 M = 0·5 N. Dissolve 65 g and make the soln up to 1 litre.

Bromine water. Br_2. F.W.=159·82 approx 0·2 M. Add 5 ml of bromine for every 100 ml of water. Shake the mixture, keep it in a dark bottle and decant off the soln as required.

Calcium chloride. $CaCl_2.6H_2O$. F.W. = 219·08 0·25 M = 0·5 N. Dissolve 55 g of crystals or 27 g of anhydrous salt in water, Make up to 1 litre.

Calcium hypochlorite (bleaching powder). Shake 125 g of solid vigorously with 1 litre of water, allow to stand and then filter.

Calcium nitrate. $Ca(NO_3)_2.4H_2O$. F.W. = 236·15 2 M = 4 N. Dissolve 470 g of dry crystals in water. Make up to 1 litre.

Calcium sulphate. $CaSO_4.2H_2O$. F.W. = 172·17 saturated = approx 0·03 M = 0·06 N. Shake 3 g of powdered salt in 1 litre of water. Stand the soln for a day and filter.

Chlorine water. Cl_2. F.W. = 70·91 sat soln = approx 0·09 M. Saturate water with chlorine, simply prepared by dropping conc hydrochloric acid on potassium permanganate crystals. Keep in a dark bottle.

Chlorine in carbon tetrachloride. Pass chlorine, washed by bubbling through water, through carbon tetrachloride to saturate it.

Cobalt nitrate. $Co(NO_3)_2.6H_2O$. F.W. = 291·04 0·01 M = 0·02 N. Dissolve 2·9 g and make up to 1 litre in water. Also used as 10% solution containing 100 g per litre and as a 5% soln by dissolving 50 g in water, adding 200 ml of glacial acetic acid and making up to 1 litre.

Cupric sulphate. $CuSO_4.5H_2O$. F.W. = 249·68 0·5 M = N. Dissolve 125 g of crystals. Make up to 1 litre in water. 0·005 M soln (1·2 g per litre) is used in the mercuri-thiocyanate test for zinc.

Cuprous chloride (acid). CuCl or Cu_2Cl_2. F.W. = 98·99 (first formula). Several methods are available.

Method 1. Shake vigorously 100 g of cupric chloride crystals, $CuCl_2.2H_2O$ and 50 g of copper turnings in 200 ml of conc hydrochloric acid. Allow to stand with occasional shaking until the soln is nearly colourless.

Method 2. (Winkler's method). Add with constant stirring 86 g of cupric oxide and 17 g of reduced copper (made by passing coal gas over heated cupric oxide) to 1 litre of 8 M hydrochloric acid. Shake the solution and suspend in it a long coil of copper wire reaching to the bottom of the solution. The soln is ready for use when it becomes colourless.

Cuprous chloride (alkaline). Neutralize the acid soln with ammonia soln until it smells of ammonia. Metallic copper must always be present in the soln.

Ferric chloride. $FeCl_3.6H_2O$. F.W. = 270·30 0·5 M. Dissolve 135 g in water, add 11–12 ml of conc hydrochloric acid and make the soln up to 1 litre.

Ferrous sulphate. $FeSO_4.7H_2O$. F.W. = 278·02 0·5 M. Dissolve 140 g in water containing 10 ml of conc sulphuric acid and make up to 1 litre.

Hydrogen Peroxide. Use 10 volume solution.

Hydrogen sulphide soln. H_2S. F.W. = 34·08 sat soln = 0·1 M approx. Pass hydrogen sulphide, washed by bubbling it through water, into water to form a sat soln.

Iodine solution. I_2. F.W. = 253·81 0·05 M = 0·1 N. Dissolve 12·7 g of crystals in water containing 20 g of potassium iodide. Make up to 1 litre.

Iodine, tincture of. Dissolve 50 g of potassium iodide in approx 50 ml of water, add 70 g of iodine. Make up to 1 litre with ethanol.

Lead acetate. $(CH_3COO)_2Pb.3H_2O$. F.W. = 379·24 0·1 M = 0·2 N. Dissolve 38 g of crystals in water. Make up to 1 litre.

Lead nitrate. $Pb(NO_3)_2$. F.W. = 331·20 0·1 M = 0·2 N. Dissolve 33 g in water. Make up to 1 litre.

Lime water. See calcium hydroxide.

Magnesia mixture (for phosphate and arsenate test). Dissolve 55 g of magnesium chloride crystals and 100 g of ammonium chloride in water, add 50 ml of 0·88 ammonia soln and make up to 1 litre. The ammonia solution may be omitted until just before use to avoid the turbidity which appears on standing.

Magnesium nitrate reagent (for phosphate and arsenate test). Dissolve 128 g of magnesium nitrate crystals and 240 g of ammonium nitrate in water, add 20 ml of 0·88 ammonia soln. Dilute to 1 litre.

Magnesium sulphate. $MgSO_4.7H_2O$. F.W. = 246·48 0·25 M. Dissolve 62 g of crystals and make up to 1 litre.

Magnesium uranyl acetate. $(CH_3COO)_2Mg.3(CH_3COO)_2.UO_2$.

Method 1. Dissolve 12 g of the solid in 200 ml of M acetic acid.

Method 2. Add 6 g of magnesium acetate to 100 ml of M acetic acid and boil until clear. Repeat with 6 g of uranyl acetate in 100 ml of M acetic acid. Mix the two hot solutions, stand for 24 hours and filter if necessary.

Manganous chloride. $MnCl_2.4H_2O$. F.W. = 197·91 0·5 M. Dissolve 99 g in water. Make up to 1 litre.

Mercuric chloride. $HgCl_2$. F.W. = 271·50 0·1 M = 0·2 N. Dissolve 27 g in water. Make up to 1 litre. A sat soln contains approx 40 g per litre.

Mercurous nitrate. $HgNO_3.H_2O$ or $Hg_2(NO_3)_2.2H_2O$. F.W. = 280·61 0·2 M (first formula). Dissolve 50 g in 40 ml of conc nitric acid. Make up to 1 litre.

Nessler's reagent (for ammonia). Dissolve 3·5 g of potassium iodide in water and add the soln to 1·25 g of finely powdered mercuric chloride in a mortar. Grind the two together until the mercuric chloride has dissolved. Wash the soln out of the mortar and add enough sat mercuric chloride soln to produce a faint ppt. Dissolve 12 g of sodium hydroxide in the soln and dilute it to 100 ml. Keep in a dark coloured bottle.

Potassium antimonate. $KSbO_3.3H_2O$. F.W. = 262·90. Either boil the solid in 2 M potassium hydroxide soln until no more dissolves and filter the solution, or boil 22 g of solid in 1 litre of water, cool rapidly and add 30 ml of 2 M potassium hydroxide.

Potassium chromate. K_2CrO_4. F.W. = 194·20 0·1 M = 0·2 N. Dissolve 20 g in water. Make up to 1 litre.

Potassium cyanide. KCN. F.W. = 65·12 0·5 M = 0·5 N. Dissolve 3·3 g in 100 ml of water. The soln deteriorates and is extremely poisonous; only small quantities should be made up.

Potassium ferricyanide. $K_3Fe(CN)_6$. F.W. = 329·26 0·133 M = 0·5 N. Dissolve 55 g in water. Make up to 1 litre. It is common, because the soln deteriorates, to wash a small crystal once or twice with water and then dissolve it for use as required.

Potassium ferrocyanide. $K_4Fe(CN)_6.3H_2O$. F.W. = 422·41 0·1 M. Dissolve 40 g and make up to 1 litre in water.

Potassium iodide. KI. F.W. = 166·01 0·2 M = 0·2 N. Dissolve 33 g in water. Make up to 1 litre.

Potassium permanganate. $KMnO_4$. F.W. = 158·04 0·02 M = 0·1 N. Dissolve 3·2 g in water by heating and make up to 1 litre. Filter through glass wool or asbestos.

Potassium oxalate. $(COOK)_2.H_2O$. F.W. = 184·24 0·5 M = 1 N. Dissolve 138 g in water. Make up to 1 litre.

Potassium thiocyanate. KSCN. F.W. = 97·18 0·1 M = 0·1 N. Dissolve 10 g in water. Make up to 1 litre.

Silver nitrate. $AgNO_3$. F.W. = 169·88 0·1 M = 0·1 N. Dissolve 17 g. in water. Make up to 1 litre. Keep in brown bottles.

Silver nitrate (alcoholic). Dissolve 1·7 g in 100 ml of ethanol. Keep in brown bottle.

Silver sulphate. Ag_2SO_4. F.W. = 311·80 0·025 M = 0·05 N approx. Shake 8 g in 1 litre of water to form almost sat soln.

Sodium acetate. $CH_3COONa.3H_2O$. F.W. = 136·08 M = N. Dissolve 136 g in water. Make up to 1 litre. 3 M is also used.

Sodium carbonate. $Na_2CO_3.10H_2O$. F.W. = 286·14 M = 2 N. Dissolve 286 g of hydrated salt, or 106 g of anhydrous salt, in water (warm for the anhydrous salt). Make up to 1 litre. Keep in a bottle with rubber or plastic stopper or cork.

Sodium chloride (normal saline). NaCl. Dissolve 4·5 g of salt in 500 ml of freshly boiled distilled water and transfer to a bottle when cool.

Sodium cobaltinitrite. $Na_3Co(NO_2)_6$. F.W. = 403·98 0·16 M = 0·32 N. Dissolve 68 g of the solid in water. Make up to 1 litre.

Alternative Method. Dissolve 30 g of cobalt nitrate in 200 ml of water and 240 g of sodium nitrite in 200 ml of water, mix the solns and pour in 60 ml of glacial acetic acid. Dilute to 1 litre. The soln is not very stable and lasts only three or four weeks. Filtration may be necessary after allowing the soln to stand.

Sodium hypochlorite. NaOCl. F.W. = 74·44 approx M = 2 N. The soln as purchased contains 10–14% of hypochlorite and is approx 2 M. Dilute with an equal volume of water, or 100 ml to 1 litre.

Alternative Method. Bubble chlorine through 200 ml of 12·5 M sodium hydroxide to which has been added 500 g of crushed ice. When the weight has increased by approx 71 g dilute to 1 litre.

Sodium hypobromite. NaOBr. F.W. = 118·9 2 M = 4 N. Stand 900 ml of 5 M sodium hydroxide soln (in a beaker) in a mixture of ice and water in a trough in a fume cupboard. Add with stirring, 2 or 3 ml at a time, 50 ml of bromine. The soln is unstable and it is probably preferable to prepare smaller amounts.

Sodium nitroprusside. $Na_2Fe(CN)_5.NO.H_2O$. F.W. = 297·97 0·03 M. Dissolve 1 g in 100 ml of water. It should be freshly prepared.

Sodium perchlorate. $NaClO_4$. F.W. = 122·44 0·2 M. Dissolve 20 g of solid in 50 ml of water and add 50 ml of ethanol.

Sodium phosphate (disodium hydrogen phosphate). $Na_2HPO_4.12H_2O$. F.W. = 358·14 0·33 M = N. Dissolve 120 g in water. Make up to 1 litre.

Sodium sulphide. $Na_2S.9H_2O$. F.W. = 240·18 0·5 M = N. Dissolve 120 g of the solid in water. Make up to 1 litre.

Alternative Method. Saturate 500 ml of M sodium hydroxide soln with hydrogen sulphide and add 500 ml of M sodium hydroxide soln.

Stannous chloride. $SnCl_2.2H_2O$. F.W. = 225·63 0·5 M = N. Dissolve, by heating if necessary, 113 g of the solid in 200 ml of conc hydrochloric acid, add several pieces of metallic tin and make the soln up to 1 litre.

Zinc nitrate. $Zn(NO_3)_2.6H_2O$. F.W. = 297·47 0·25 M = 0·5 N. Dissolve 74 g of crystals in water. Make up to 1 litre.

Zinc uranyl acetate. $(CH_3COO)_2Zn+(CH_3COO)_2.UO_2$. Dissolve 10 g of uranyl acetate and 30 g of zinc acetate in 12 ml of 5 M acetic acid, dilute to 100 ml, add about 0·1 g of sodium chloride and filter after 24 hours.

Zirconium nitrate. $ZrO(NO_3).2H_2O$. F.W. = 267·26. (*a*) For fluoride test: dissolve 1 g of solid in 200 ml of conc hydrochloric acid and dilute to 1 litre. (*b*) For phosphate separation: add 10 g of solid to 100 ml of M nitric acid and boil the mixture with stirring. Allow to stand for 24 hours and filter through glass wool.

5

Indicators

ACID-BASE OR pH INDICATORS

Indicators suitable for acid-base titrations and for the measurement of the pH of a solution are numerous and a selection of those in common use is given. They all undergo a change in colour when the pH of the solution in which they are contained alters; the change in pH required to produce the complete change in colour of the indicator is called its *pH range* or *pH interval*.

The normal colour change of an indicator may be altered by two methods. The addition of a dye to the solution will cause some of the light which would normally be transmitted to be absorbed. For example, the colour change from yellow-orange to pink given by the indicator methyl orange can be altered to a change from green to violet by adding the dye xylene cyanol, which is blue. Such indicators are said to be *screened*.

The other method is to use a mixture of two or more indicators in the same solution. Where such a mixture gives a series of different colours for a range of different pH values it is called a *universal indicator*. The majority of acid-base indicators are usually dissolved in ethanol (ethyl alcohol), although the solution so formed is often diluted with water. Water-soluble forms of some indicators, usually their sodium salts, can be made or purchased.

The table shows single or screened indicators, the pH range, the colour change, the weight per litre of solution and the volume of ethanol in which it should be dissolved, water being added to make the volume up to 1 litre. They cover the pH range from 0 to 13. A second table lists others in alphabetical order.

Indicator	pH range	Colour change	Weight in g	Ethanol volume in ml
Methyl violet	0·1–2·0	yellow–violet	0·5	0
Cresol red	0·2–1·8	red–yellow	0·5	200
m-Cresol purple	1·0–2·6	red–yellow	0·4	200
Thymol blue	1·2–2·8	red–yellow	0·4	200
Dimethyl yellow	2·9–4·0	red–yellow	0·4	900
Quinaldine red	1·4–3·2	colourless–pink	0·4	1000
Bromophenol blue	2·8–4·6	yellow–blue violet	0·4	200
Methyl orange	2·9–4·6	red–yellow	0·4	200
Methyl orange-xylene cyanol	2·9–4·6	violet–green	3·6	200
Congo red	3·0–5·0	blue–red	0·2	200
Bromocresol green	3·6–5·2	yellow–blue	0·4	200
Methyl red	4·2–6·3	red–yellow	0·2	300
Chlorophenol red	4·8–6·4	yellow–violet red	0·4	200
Bromocresol purple	5·2–6·8	yellow–purple	0·4	200
Bromothymol blue	6·0–7·6	yellow–blue	0·4	200
Phenol red	6·8–8·4	yellow–red	0·2	200
Cresol red	7·2–8·8	yellow purple	0·2	200
m-Cresol purple	7·6–9·2	yellow–violet blue	as above	
Thymol blue	8·0–9·6	yellow–violet	0·4	200
Phenolphthalein	8·2–10	colourless–violet red	1·0	600
Phenolthymolphthalein	8·3–11	colourless–violet	1·0	600
Thymolphthalein	9·3–10·5	colourless–blue	2·0	1000
Brilliant orange	10·5–12·0	yellow–red	0·4	200
Alizarin	11·0–13·0	pink–violet	1·0	0
Tropaeolin 000	11·0–13·0	yellow–red	2·0	600
Tropaeolin 0	11·1–12·7	yellow–orange	0·1	600
Titan yellow	12·0–13·0	yellow–red	1·0	500

SCREENED INDICATORS

Indicator	pH range	Colour change	Preparation of 1 litre
Dimethyl yellow Methylene blue	2·9–4·0	pink–yellow green	1 g dimethyl yellow + 0·5 g of methylene blue in ethanol
Methyl orange Xylene-cyanol	2·9–4·6	mauve–green	1 g methyl orange + 2·6 g xylene-cyanol in water
Methyl orange Aniline green	2·9–4·6	violet–green	0·5 g methyl orange + 0·5 g analine green in water
Methyl red Methylene blue	4·2–6·3	mauve–green	1 g methyl red + 0·5 g methylene blue in ethanol
Chlorophenol red Aniline blue	4·8–6·4	green–violet	0·5 g chlorophenol blue (Na) 0·5 methylene blue in water
Neutral red Methylene blue	6·8–8·0	blue violet–green	0·5 g neutral red + 0·5 g methylene blue in ethanol
Phenolphthalein Methyl green	8·0–9·8	green–violet	0·33 g phenolphthalein + 0·66 g methyl green in ethanol

MIXED INDICATORS

Indicator	*Midway pH*	*Colour change*	*Preparation of 1 litre*
Methyl orange Bromocresol green (Na salt)	4·3	orange–blue green	0·66 g methyl orange + 0·33 g bromocresol green in water
Methyl red Bromocresol green	5·1	wine red–green	0·4 g methyl red + 0·6 g bromocresol green in ethanol
Chlorophenol red Bromocresol green (use Na salts)	6·1	yellow green–violet	0·5 g of each in water
Bromothymol blue Bromocresol purple	6·7	yellow–blue violet	0·5 g of each Na salt in water
Bromothymol blue Azolitmin	6·9	violet–blue	0·66 g bromothymol blue (Na salt) + 0·33 g azolitmin in water
Neutral red Bromothymol blue	7·2	rose–green	0·5 g of each indicator in ethanol
Cyanine Phenol red	7·3	yellow–violet	0·66 g cyanine + 0·33 g phenol red in 500 ml ethanol and 500 ml of water
Phenol red Bromothymol blue (Na salts)	7·5	yellow–violet	0·5 g of each in ethanol or 0·5 g of water-soluble form of each in water
Cresol red Naphtholphthalein	8·3	pale pink–violet	0·66 g naphtholphthalein + 0·33 g cresol red in ethanol
Cresol red Thymol blue (use Na salts)	8·3	yellow–violet	0·25 g cresol red + 0·75 g thymol blue (both Na salts) in water
Thymol blue Phenolphthalein	9·0	yellow–violet	0·75 g phenolphthalein + 0·25 g thymol blue in 500 ml of ethanol + 500 ml of water
Thymolphthalein Phenolphthalein	9·9	colourless–violet	0·5 g of each in ethanol
Thymolphthalein Alizarin yellow	10·2	yellow–violet	0·66 g thymolphthalein + 0·33 g alizarin yellow in ethanol

WATER-SOLUBLE INDICATORS

In order to avoid using ethanol as a solvent, some indicators can be made water-soluble. They can be obtained commercially in tubes which each contain enough solid to make a litre of solution. They are referred to in the lists above as '(Na)' or '(Na salts)', and can be made by the Clark and Lubs method using sodium hydroxide solution. The quantities of 0·01 M solution needed for 0·1 g of indicator are given in the table.

Indicator	*ml of 0·01M NaOH*	*pH range*
Cresol red	26·2	0·2–1·8
m-Cresol purple	26·2	1·0–2·6
Bromophenol blue	14·9	2·8–4·6
Bromocresol green	14·3	3·6–5·2
Chlorophenol red	23·6	4·8–6·4
Bromocresol purple	18·5	5·2–6·8
Bromophenol red	19·5	5·2–7·0
Bromothymol blue	16·0	6·0–7·6
Thymol blue	21·5	8·0–9·6

For example, to make a solution of water-soluble bromothymol blue of the same concentration as the normal indicator, use 0·4 g per litre and, from the table, 64·0 ml of 0·01 M sodium hydroxide solution.

OTHER pH INDICATORS

(Weight of indicator and volume of ethanol for 1 litre)

Indicator	*pH range*	*Colour change*	*Weight of indicator g*	*Volume of ethanol ml*
Alizarin red S	4·0– 6·0 6·0–12·0	yellow–orange red orange red–violet	1·0	none
Alizarin yellow G	10·1–12·1	yellow–red orange	0·1	none
Alizarin yellow GG	10·0–12·0	colourless–yellow	0·1	none
Alkali blue 6B	alcoholic titrations	blue–red	1·0	1000
Azolitmin	5·0–8·0	red–blue	1·0	none
Brilliant yellow	6·4–9·4	yellow–red orange	0·4	200
Bromochlorophenol blue	3·0–4·2	yellow–blue	0·4	200
Bromophenol red	5·2–7·0	yellow–red purple	0·4	200
Cochineal	5·0–6·0	orange–deep red violet	in solution	
o-Cresolphthalein	8·2–9·8	colourless–violet red	0·2	800

Indicator	*pH range*	*Colour change*	*Weight in g*	*Volume of ethanol in ml*
Cyanine (quinoline blue)	6·6–8·6	colourless–blue	1·0	1000
Cyanine acid blue	10·5–12·0	blue–red	0·4	500
2:4-Dinitrophenol	2·8–4·4	colourless–yellow	1·0	500
2:5-Dinitrophenol	4·0–5·6	colourless–yellow	1·0	500
2:6-Dinitrophenol	2·2–4·4	colourless–yellow	1·0	500
Diphenol purple	7·0–8·6	yellow–purple	0·4	200
Ethyl bis(2:4-dinitrophenol) acctate	7·4–9·1	colourless–deep blue	sat	500 ethanol 500 acetone
Ethyl orange	3·0–4·5	red–orange	0·4	200
Gallein (pyrogallolphthalein)	3·4–6·8	pale orange	1·0	800
Haematoxylin	0·0–1·0 6·0–11·0	red–yellow orange–red violet	1·0	1000
Indigo carmine	11·6–14	blue–yellow	1·0	500
Lacmoid	4·0–6·0	orange red–violet	5·0	1000
Litmus	5·0–8·0	red–blue	see below	
Martius yellow (Aniline yellow)	2·0–3·2	colourless–yellow	0·5	1000
Meta-methyl red	2·0–4·0	red–yellow	0·4	200
Metanil yellow	1·2–3·2	red–yellow	0·5	1000
p-Naphthol-benzein (Naphthol-benzein)	8·5–9·8	yellow–green	1·0	700
α-Naphthol-phthalein	7·3–8·7	yellow–blue	0·4	800
Neutral red	6·8–8·0	red–orange	1·0	600
m-Nitrophenol	6·8–8·4	colourless–yellow	1·0	none
p-Nitrophenol	5·0–7·0	colourless–yellow	1·0	none
Para-methyl red	1·2–3·4	red–orange	0·4	200
Phenol violet	8·0–10	yellow–blue–violet	0·2	200
N-phenyl-1-naphthylamine-azo-benzene-p-sulphonic acid	3·0–5·0	violet–red orange	0·5	200
Quinizarin-sulphonic acid (rufianic acid)	8·5–10 10·5–13	orange–purple purple–blue	0·4	none
Phenol tetrachlorophthalein	8·2–9·4	colourless–pink	0·4	200
Poirrier blue C4B	11·0–13·0	blue–red	1·0	none
Resazurin	5·0–7·0	pink–violet	0·05	freshly prepared in water
Rosolic acid (aurin or corallin)	6·9–8·0	yellow brown–red	5·0	500
Tetrabromophenol blue	3·0–5·0	yellow–blue	0·4	200

Indicator	*pH range*	*Colour change*	*Weight of indicator g*	*Volume of ethanol ml*
Thymol violet	9·0–13	yellow–green–violet	0·4	200
1:3:5-Trinitrobenzene	11·5–14	colourless–orange	1·0	1000
2:4:6-Trinitrobenzoic acid	12·0–13·4	colourless–orange–red	1·0	none
p-Xylenol blue	1·2–2·8	red–yellow	0·4	500
(p-Xylenolsulphon phthalein)	7·8–9·4	yellow–blue		

LUMINESCENT INDICATORS

These are substances which mark the equivalence point of an acid-base titration if the solution is placed in ultraviolet light, with the rest of the room darkened. The luminosity which they show in ultraviolet light changes or changes colour as the pH of the solution changes over a short range. They are useful when the colour or turbidity of a titration solution would interfere with normal indicators. They are made by dissolving 0·1 g in 100 ml of ethanol.

Indicator	*pH range*	*Colour change*
Eosin	2·5–4·5	colourless–green
Chromotropic acid	3·5–4·5	colourless–blue
Fluorescein	4·0–6·0	colourless–green
Dichlorofluorescein	4·0–6·5	colourless–green
Acridine	5·2–6·6	green–violet
β-Naphthol	8·5–9·5	colourless–blue
Quinine	3·0–5·0	blue–violet
	9·5–10·0	violet–colourless

Manufacturers produce a series of mixed indicators of various pH ranges for particular purposes; the pH range is often indicated in the trade name.

UNIVERSAL INDICATORS

Yamada's universal indicator (*Chem. Abstr.*, 1929, 23, 752, 2120). 0·025 g of thymol blue, 0·0625 g of methyl red, 0·5 g of phenolphthalein and 0·25 g of bromothymol blue in 500 ml of ethanol made up to 1 litre with water. Either before or after dilution, drops of approx 0·05 M sodium hydroxide are added until the solution is green in colour.

Colour	red	orange	yellow	green	blue	dark blue	violet
pH	4	5	6	7	8	9	10

van Urk's universal indicator (van Urk, *Chem Abstr.*, 1934, 28, 2258). 0·35 g of Tropaeolin 00, 0·5 g of methyl orange, 0·4 g of methyl red, 2·0 g of bromothymol blue, 2·5 g of α-napthol phthalein 2·0 g of o-cresol phthalein, 2·5 g of phenolphthalein and 0·75 g of alizarin yellow R dissolved in 700 ml of ethanol and made up to 1 litre with water.

Colour	orange–red	red–orange	orange	yellow–orange	orange–yellow	yellow	green–yellow	green
pH	2	3	4	5	6	6·5	7	8

Colour	blue–green	green–blue	violet–blue	violet	violet–violet red	violet–red
pH	8·5	9	9·5	10	11	12

BUFFER SOLUTIONS

Buffer solutions of known pH are obtainable commercially already made up and tested, or as concentrated solutions with directions for preparation from them of solutions of the required pH. Formulae for the preparation have been suggested by many authors and the necessary standard solutions can also be obtained commercially or made up from reagents of analytical standard of purity.

Prideaux recommends using decimolar solutions of sodium dihydrogen phosphate and disodium hydrogen phosphate for simply prepared solutions of between pH 4 and pH 9.

ml of NaH_2PO_4 soln	10	8	6	4	2	0
ml of Na_2HPO_4 soln	0	2	4	6	8	10
pH	4·0	6·0	6·5	6·75	7·2	9·0

(Prideaux, *Theory and use of Indicators,* van Nostrand 1917.)

Buffer solutions according to *Clark and Lubs.*

The two solutions are mixed in the proportions shown in the table and the mixture is diluted to 100 ml exactly.

pH	*x ml of HCl*	
1·0	48·5	25 ml of 0·2 M
1·2	32·25	potassium chloride
1·4	20·75	+
1·6	13·15	*x* ml of 0·02 M
1·8	8·3	hydrochloric acid
2·0	5·3	solutions
2·2	3·35	

pH	*x ml*	
2·2	46·70	50 ml of 0·1 M
2·4	39·60	potassium hydrogen
2·6	32·95	phthalate
2·8	26·42	*x* ml of 0·1 M
3·0	20·32	hydrochloric acid
3·2	14·70	solutions
3·4	9·90	
3·6	5·97	
3·8	2·63	

pH	*x ml*	
4·0	0·40	
4·2	3·70	50 ml of 0·1 M
4·4	7·50	potassium hydrogen
4·6	12·15	phthalate
4·8	17·70	+
5·0	23·85	*x* ml of 0·1 M
5·2	29·95	sodium hydroxide
5·4	35·45	
5·6	39·85	
5·8	43·00	
6·0	45·45	

pH	*x ml*	
6·0	5·70	
6·2	8·60	50 ml of 0·1 M
6·4	12·60	potassium
6·6	17·80	dihydrogen
6·8	23·65	orthophosphate
7·0	29·63	+
7·2	35·00	*x* ml of 0·1 M
7·4	39·50	sodium hydroxide
7·6	42·80	solutions
7·8	45·20	
8·0	46·80	

pH	*x ml*	
7·8	2·61	
8·0	3·97	50 ml of 0·1 M
8·2	5·90	boric acid in
8·4	8·50	0·1 M potassium
8·6	12·00	chloride
8·8	16·30	+
9·0	21·30	*x* ml of 0·1 M
9·2	26·70	sodium hydroxide
9·4	32·00	solutions
9·6	36·85	
9·8	40·80	
10·0	43·90	

INDICATORS FOR PRECIPITATION TITRATIONS

Potassium chromate. Dissolve either 50 g of potassium chromate crystals in water or 42 g of the chromate and 7 g of potassium dichromate and make the soln up to 1 litre. The chromate-dichromate mixture has a buffer action which improves the effectiveness of the indicator in slightly acid or alkaline soln. 1 ml of indicator soln per 50 ml of final soln is recommended for titrations using 0·1 N soln.

Adsorption Indicators

Indicator	*Colour change*	*For determination of*	*Grams per litre*	*Solvent*
Alizarin red S	yellow–pink	ferrocyanide molybdate fluoride	1·0	water
Bromophenol blue	colourless–blue	halides thiocyanate mercurous silver thallous	0·4	200 ml ethanol: 800 ml water
Chromotrope F4B	grey green–yellow green	bromide iodide	1·0	water
Chrome azurol S	pale pink–deep pink	fluorides	1·0	water
Congo red	violet–red	halides thiocyanate	1·0	water
Dibromo(R)fluorescein	yellow green–pink	chlorides borates	1·0	ethanol
Dichlorofluorescein	yellow green–pink	chloride borate bromide iodide	1·0	ethanol (or sodium salt in water)
Di-iodo(R)dimethyl (R)fluorescein	orange pink–red	iodide	1·0	ethanol
Di-iodo(R)fluorescein	yellow–pink	chloride iodide	1·0	ethanol
Diphenyl carbazone	orange pink–violet	cyanide	1·0	ethanol
Eosin (tetrabromo(R)fluorescein)	pink–red purple	bromide iodide	1·0	700 ml ethanol: 300 ml water (or sodium salt in water)
Erythrosin (tetraiodo(R) fluorescein)	pink–blue pink	iodides	1·0	700 ml ethanol: 300 ml of water

Indicator	Colour change	For determination of	Grams per litre	Solvent
Fluorescein	green–pink	halides thiocyanate sulphate	1·0	ethanol
Fluorescein sodium	as for fluorescein		1·0	water
Mercurochrome	pale orange–strong pink	thiocyanate chloride	1·0	water
Phenosafranine	red–blue	chloride bromide	1·0	water
Rhodamine 6G	orange pink–violet	silver	1·0	water
Solochrome red B	soln: colourless–orange red. ppt: blue pink–white	molybdate ferrocyanide	2·0	water
Tartrazine	soln: colourless–yellow	halides thiocyanate silver	1·0	water
Titan yellow	yellow–blue	chloride and bromide with mercurous nitrate	1·0	500 ml ethanol + 500 ml of water

Ferric alum. For Volhard's method. Dissolve 100 g of crystals in 1 litre of water and add enough 2 M nitric acid to clear the soln and prevent hydrolysis. Colour change: white ppt to orange red colouration. Use 1 ml.

STARCH INDICATOR FOR IODINE TITRATIONS

Make 10 g of soluble starch (A purity) into a smooth paste with water. Pour it into a litre of boiling water. Cool. It deteriorates on keeping. Making as required is preferable to using preservatives.

Sodium starch glycollate. Dissolve 1 g in 100 ml of hot water and cool the soln.

INDICATORS FOR REDOX (OXIDATION-REDUCTION) REACTIONS

A redox indicator is a substance which changes colour on reduction. Its colour change is readily reversed by oxidation. The complete colour change takes place for a small and specific change in the redox potential of the solution in which the indicator is placed. The potential at which the oxidized and reduced state of the indicator are equal in concentration is called the *normal*, or *transition, potential* and represented by E_0. Its value alters with the pH of the solution.

A series of such indicators can therefore, by means of their colours, be used to measure the redox potential of a chemical or biological system. When used in a redox titration, the value of E_0 for the indicator must match the potential of the solution at the point when equivalent quantities of oxidizing and reducing agent have been added. The choice of indicator depends on the reaction involved, as shown in the table. The potential at the equivalence point can be altered slightly by altering the conditions in solution.

TITRIMETRIC OR VOLUMETRIC INDICATORS

Indicator	*Approx* E_0+	*Colour change*	*Determinations*	*Preparation*
Barium diphenylamine-sulphonate	+0·83	green–violet	of ferrous ion	0·2 g/100 ml of water
3:3′-Dimethyl-naphthidene	0·80	colourless–red violet	of zinc and other metals by ferrocyanide	0·1 g in 100 ml of glacial acetic acid
3:3′-Dimethyl-naphthidene-disulphonic acid	0·71	colourless–red violet	of zinc and other metals by ferrocyanide	1 g in 100 ml of water
Diphenylamine	0·76	colourless–violet	by dichromate titrations	1 g in 100 ml of conc sulphuric acid
Diphenylbenzidene	0·76	colourless–violet	of zinc by ferrocyanide	1 g in 100 ml of conc sulphuric acid
2:2′-Dipyridil (α-α-Dipyridil)	0·97	red–faint blue	by ceric sulphate soln titrations	1·172 g+0·695 g of $FeSO_4$ crystals in 100 ml of water
Erioglaucine A (Alphazurine G)	1·00	green yellow–orange	ceric sulphate titrations	0·1 g in 100 ml of water
Erio-green B (Lissamine green)	1·01	green–orange	ceric sulphate titrations	0·1 g in 100 ml of water
p-Ethoxychrysoidine	0·76	green grey–salmon pink	of zinc by ferrocyanide	0·02 g in 100 ml of conc sulphuric acid
		red–orange yellow	of arsenic by bromate or iodate	0·1 g in 100 ml of ethanol
α-Naphtho-flavone (7:8-benzo-flavone)	0·8	yellow–brown orange	of arsenic by bromate	0·5 g in 100 ml of ethanol

Indicator	*Approx E_0*	*Colour change*	*Determinations*	*Preparation*
1:10-phenanthroline hydrate and 1:10-phenanthroline ferrous sulphate complex (Ferroin)	1·14	red–blue	of ferrous ion by ceric sulphate titrations	1·485 g in 100 ml of water 1·485 g + 0·695 g of ferrous sulphate crystals in 100 ml of water
N-phenylanthranilic acid	1·08	colourless–cherry red	of ferrous ion by ceric sulphate or dichromate	0·1 g in 5 ml of 0·1 M sodium hydroxide. Dilute to 100 ml with water
Quinoline yellow		yellow–colourless	of arsenite by hypochlorite	0·2 g in 100 ml of water
Setopaline	1·07	yellow–orange	by ceric sulphate and ferrocyanide	0·1 g in 100 ml of water
Setoglaucine	1·0	yellow green–yellow–red	reduced ferric iron by permanganate	0·1 g in 100 ml of water
Sodium-diphenylamine-sulphonate	0·85	green–red violet	of ferrous ion	0·2 g in 100 ml of water
Xylene-cyanol	1·00	green–orange	by ceric sulphate and mercuric nitrate	0·1 g in 100 ml of water

The following are irreversible indicators. Local concentrations of the oxidizing agent can therefore reduce the amount of indicator before the end-point is reached and more may need to be added.

Indicator	*Colour change*	*Determinations*	*Preparation*
Amaranth	red–colourless	titrations with hypochlorite and iodate	0·2 g in 100 ml of water
Bordeaux	pink–faint yellow green	titrations with hypochlorite	0·2 g in 100 ml of water
Brilliant Ponceaux 5R	orange–colourless	titrations by iodate	0·2 g in 100 ml of water
Naphthol blue black	green–faint pink	titrations with iodate and bromate	0·2 g in 100 ml of water

Methyl orange and Methyl red have also been used in concentrations identical to those shown under acid-base indicators.

INDICATORS FOR EDTA TITRATIONS

Indicator	*Colour change*	*Determin-tion of*	*Preparation*
Ammonium purpurate (Murexide)	salmon pink–blue	calcium	1 g+100 g NaCl finely powdered by grinding. Use 0·2 g of the powder
Murexide with Naphthol green	green–red–blue		Grind 0·2 g+0·5 g Naphthol blue+100 g NaCl. Use 0·2 g of powder
Bromopyrogallol red and Pyrogallol red	blue–red or red–yellow	bismuth cobalt nickel etc.	0·05 g in 50 ml of ethanol+50 ml of water
Calcon		calcium	as powder made according to the Murexide formula or 1 g in 100 ml of ethanol containing 0·8 g Na_2CO_3
Catechol violet (Pyrocatechol violet)	blue–yellow or red–yellow	bismuth nickel thorium	0·1 g in 100 ml of water
Gallocyanine	blue–red	gallium	1 g in 100 ml of glacial acetic acid
Phthalein purple (Phthalein complex)	red–grey or green	calcium barium strontium sulphate	0·18 g+0·02 g naphthol green in 100 ml of water made just alkaline with ammonia
Solochrome (or Eriochrome) black T	red–blue	magnesium calcium zinc, lead manganese	as Murexide or 1 g in 100 ml of ethanol
Solochrome black 6B	red–blue or to green if screened with dimethyl yellow	as for Solochrome black T	0·5 g in 100 ml of ethanol or 6 drops of this+2 of 0·25 g of dimethyl yellow in 100 ml of ethanol
Sulphosalicylic acid	green blue–pale yellow	iron	2 g in 100 ml of water
Tiron	green blue–pale yellow or colourless	iron	2·0 g in 100 ml of water

Indicator	*Colour change*	*Determination*	*Preparation*
Variamin blue B	blue–faint yellow	ferric iron	1·0 g in 100 ml of water
Xylenel orange	red purple–lemon yellow	cadmium, lead mercury, zinc bismuth, etc.	0·1 g in 100 ml of water or in 50 ml of ethanol + 50 ml of water
PAN 1-(2-pyridolazo)-2-naphthol	red or pink–yellow	cadmium copper	0·5 g in 100 ml of ethanol

6

Organic Reagents and others used in Qualitative Analysis

Since these reagents are used in spot tests or in qualitative tests using semi-micro quantities, instructions are given for the preparation of 100 ml of soln. The chief ions for which the reagent is used are listed. Some tests can be used in the quantitative estimation of the ions listed.

Alizarin. Reagent for aluminium. A saturated soln in ethanol.

Alizarin S (sodium alizarin sulphonate). Reagent for aluminium. 0·1 g per 100 ml of water.

Aluminon (the ammonium salt of aurin tricarboxylic acid). Reagent for aluminium. 0·1 g per 100 ml of water.

1-Amino-4-hydroxy-anthroquinone. Reagent for lithium. 0·1 g in 100 ml of ethanol.

Aniline sulphate. Reagent for chlorate. 0·1 g of the solid in 100 ml of water or 1 ml of aniline in 1 litre of water and add conc sulphuric acid to dissolve it.

Benzidene. Reagent for lead, ferricyanide and persulphate. For Lead: 0·05 g in ml of 1·5 M acetic acid. For ferricyanide: sat soln in cold 2 M acetic acid. For persulphate: 2 g in 100 ml of 2 M acetic acid. There are hazards in handling benzidene.

Benzidene-cuprous acetate reagent. For cyanide. Dilute 47·5 ml of a sat soln of benzidene in glacial acetic acid to 100 ml. Dissolve 0·286 g of cuprous acetate in 100 ml of water. Use five parts of the first soln to one part by volume of the second.

Benzidene hydrochloride. Reagent for sulphate. 0·8 g dissolved in 12 ml of M hydrochloric acid and made up to 100 ml with water.

Benzoin α-oxime. See Cupron.

S-benzyl-thiuronium chloride. Reagent for nitrogen, sulphur and chlorine. 10 g per 100 ml of water.

Brucine. Reagent for nitrate and bismuth. 0·1 g in 100 ml of ethanol.

Cacotheline. Reagent for stannous tin. 0·25 g in 100 ml of water. Deteriorates on standing.

Cadion 2B (4-nitro-naphthalene-diazoamino-azobenzene). Reagent for cadmium. 0·02 g in 100 ml of ethanol, with 1 ml of 2 M potassium hydroxide.

Carmine. Reagent for boron. 0·5 g in 100 ml of ethanol.

Catechol violet. Reagent for aluminium, bismuth, tin, vanadium and titanium. 0·1 g in 100 ml of water.

Chloride reagent. Test for chlorides. Dissolve 1·7 g of silver nitrate in water, add 25 g of potassium nitrate and 17 ml of 0·88 ammonia soln and make up to 1 litre.

Chrome azurol S. Reagent for beryllium. 0·1 g in 100 ml of water.

Chromotrope 2B (sodium salt of chromotropic acid). Reagent for boron, chromium, nitrates, titanium and formaldehyde. 0·005 g in 100 ml of conc sulphuric acid.

Cinchonine. Reagent for bismuth. Dissolve 2·5 g in 100 ml of 0·25 M nitric acid soln.

o-Cresol phthalein complexone (phthalein purple). Reagent for calcium, strontium, magnesium and barium. It is usually used with naphthol green as a screening dye. 0·18 g of solid with 0·01 g of the dye in 100 ml of water containing a little ammonia.

Cupferron (copperone). Reagent for aluminium, bismuth, iron, molybdenum, tin, tungsten, uranium, vanadium and zirconium. 2 g in 100 ml of water containing dissolved in it a small piece of ammonium carbonate. The soln lasts about a week.

Cupron (α-benzoin oxime). Reagent for copper, molybdenum and tungsten. 5 g in 100 ml of ethanol.

Curcumin. Reagent for boron. 0·1 g in 100 ml of ethanol.

o-Dianisidene. Reagent for chromium gold and phosphate. 2 g in 100 ml of 6 M hydrochloric acid.

1:4-Dihydroxy-anthraquinone. See quinizarin.

1:2-Dihydroxybenzene-3:5-disulphonic acid (disodium salt). Reagent for cerium, iron, molybdenum and titanium. 0·5 g in 100 ml of water.

4-Dimethyl-amine-azobenzene-4′-arsonic acid. Reagent for zirconium. 1 g in 5 ml of conc hydrochloric acid made up to 100 ml with ethanol.

p-Dimethyl-amino-benzylidene-rhodanine. Reagent for gold, mercury and silver. 0·03 g in 100 ml of ethanol or acetone.

Dimethyl-glyoxime. Reagent for bismuth, cobalt, nickel and silver. 1 g in 100 ml of ethanol.

Di-β-naphthol. Reagent for tartrates. 0·05 g in 100 ml of conc sulphuric acid.

Dinitro-diphenyl carbazide. Reagent for cadmium. 0·1 g in 100 ml of ethanol

2:4-Dinitro-1-naphthol-7-sulphonic acid (flavionic acid). Reagent for caesium. 0·5 g in 100 ml of formic acid.

2:4-Dinitro-1-naphthol-7-sulphonic acid sodium salt (sodium flavionate). Reagent for potassium. 2 g in 100 ml of water.

Diphenylamine. Reagent for nitrate. 0·5 g in 100 ml of conc sulphuric acid.

Diphenyl-benzidene. Reagent for nitrate and nitrite. 0·02 g in 100 ml of conc sulphuric acid free from nitric acid.

Diphenyl carbazide. Reagent for arsenate, cadmium, chromium, mercury. 1 g in either 100 ml of ethanol or in 10 ml of acetic acid (glacial) diluted to 100 ml with ethanol.

Diphenyl carbazone. Reagent for germanium and mercury. 1 g in 100 ml of ethanol.

Dipicrylamine. (Hexanitro-diphenylamine reagent for potassium). 1 g dissolved in 100 ml of 0·05 M sodium carbonate by heating and filtered on cooling.

2:2′-Dipyridyl. Reagent for cadmium. 0·2 g with 0·146 g of ferrous sulphate crystals in 50 ml of water. Dissolve 10 g of potassium iodide by shaking and filter.

Dithiol (4-methyl-1:2-dimercapto-benzene). Reagent for tin. 0·2 g in 100 ml of 0·25 M sodium hydroxide soln.

Dithio-oxamide (rubeanic acid). Reagent for cobalt, iron, copper, nickel. 0·5 g in 100 ml of ethanol.

Dithizone (diphenyl-thiocarbazone). Reagent for cadmium, gold, indium, mercury, silver, zinc. 0·005 g in 100 ml of carbon tetrachloride or chloroform.

Ethyl-ammonium ethyl-dithiocarbamate (Emdite). Reagent for use as an alternative to hydrogen sulphide and used for the precipitation of antimony, arsenic, cadmium, cobalt, iron, lead, manganese, mercury, nickel, silver and tin. 50% soln by weight.

Ethylene-diamine. Reagent for mercury. Add ethylene diamine solution in water to 0·1 M copper sulphate soln until the blue-violet colour does not increase in intensity.

Ferric periodate. Reagent for lithium. 2 g of potassium periodate in 10 ml of 2 M potassium hydroxide soln, dilute to 50 ml approx with water, add 3 ml of 10% ferric chloride soln and dilute to 100 ml with 2 M potassium hydroxide soln.

Fluorescein. Reagent for bromide. Sat soln in 50 ml of water and 50 ml of ethanol.

Formaldoxime. Reagent for manganese. 2·5 g in 100 ml of water, or dissolve 4 g of hydroxylamine hydrochloride in 2 ml of 40% formalin soln and make up to 100 ml with water.

α-Furil-dioxime. Reagent for nickel. 10 g in 100 ml of ethanol.

Gallic acid. Reagent for cerium. 0·02 g in 100 ml of water.

Gallocyanine. Reagent for lead. 1 g in 100 ml of water, or glacial acetic acid.

8-Hydroxy-7-iodo-quinoline-5-sulphonic acid (Ferron). Reagent for aluminium and iron. 0·2 g in 100 ml of water.

4-Hydroxy-3-nitro-phenyl-arsenic acid. Reagent for calcium, cadmium and tin. Sat soln in water.

8-Hydroxy-quinaldine. Reagent for lead, magnesium, titanium and zinc. 0·2 g in 100 ml of water.

8-Hydroxy-quinoline (Oxine). Reagent for aluminium, cadmium, cobalt, indium, molybdenum, tin, titanium and zinc. 0·2 g in 100 ml of water.

Indigo. Reagent for hydrosulphites. 0·4 g of indigo allowed to stand in 5 ml of conc sulphuric acid after warming. Some hours later pour the soln into 95 ml of water.

Indole. Reagent for nitrite. 0·02 g in 100 ml of water.

Magenta (fuchsin). Reagent for sulphites. 0·02 g in 100 ml of water.

2-Mercapto-benzo-thiazole. Reagent for bismuth, cadmium and lead. 2 g in 100 ml of 0·1 M potassium hydroxide soln.

Methyl violet. Reagent for antimony. 0·1 g in 100 ml of water.

α-Naphthylamine. Reagent for gold. 0·5 g in 100 ml of ethanol.

Nickel ethylene-diamine. Reagent for thiosulphate. Add ethylene-diamine to 0·5 M soln of nickel nitrate to give a violet blue colour.

Nioxime. Reagent for nickel and others. 0·8 g in 100 ml of water.

p-Nitrobenzene-diazo-amino-azo-benzene. See Cadion 2B.

p-Nitrobenzene-diazonium chloride (Riegler's reagent). Reagent for ammonium. Add to 0·5 g of p-nitraniline dissolved in a soln of 1 ml of conc hydrochloric acid in 9 ml of water, enough water to make the solution up to 90 ml. Cool in ice and add 10 ml of a soln of sodium nitrite containing 0·5 g of dissolved crystals, with constant shaking.

Nitron. Reagent for nitrate, perchlorate, rhenate, tetra-fluorborate and tungsten. 10 g in 100 ml of 0·1 M acetic acid.

4-(p-Nitro-phenyl-azo)-chromotropic acid sodium salt. See chromotrope 2B.

4-(p-Nitro-phenyl-azo)-1-naphthol (Magneson II). Also called p-nitrobenzene-azo-α-napthol. Reagent for magnesium. 0·001 g in 100 ml of M sodium hydroxide soln.

4-(p-Nitro-phenyl-azo)-orcinol. Reagent for beryllium. 0·025 g in 100 ml of M sodium hydroxide soln.

4-(p-Nitro-phenyl-azo)-resorcinol (Magneson I). Reagent for magnesium. 0·01 g in 100 ml of M sodium hydroxide soln.

The alternative names for the last two reagents are exactly comparable to the alternative for Magneson II.

1-Nitroso-2-naphthol (α-nitroso-β-napthol). Reagent for cobalt iron, nickel and palladium. 1 g in 50 ml of water with 50 ml of glacial acetic acid, or in 100 ml of ethanol or acetone.

Nitroso-R-salt (1-nitroso-2-hydroxynaphthalene-3:6-disulphonate sodium salt). Reagent for cobalt and iron. 1 g in 100 ml of water.

Pentahydroxy-flavone (Morin). Reagent for aluminium, beryllium, boron, uranium and zirconium. 5 g in 100 ml of ethanol.

Phenazone. Reagent for antimony and cobalt. 1 g in 100 ml of water.

Phenyl-arsonic acid. Reagent for bismuth, tin and zirconium. 10 g in 100 ml of water.

Phenyl-fluorone (9-phenyl-2:3:7-trihydroxy-6-fluorone). Reagent for germanium.

Phenyl-hydrazine. Reagent for molybdate. 1·5 g of phenyl hydrazine in 50 ml of glacial acetic acid and 50 ml of water.

Phosphomolybdic acid. Reagent for antimony. 5 g in 100 ml of water.

Picrolonic acid. Reagent for calcium, lead and thorium. Sat soln in water.

Potassium periodate. Reagent for manganese. Sat soln in water.

Precipitation reagent (for anion analysis after boiling the solid with sodium carbonate). Dissolve 52 g of calcium chloride (hydrated) and 61 g of barium chloride crystals in water and make up to 1 litre.

1-(2-Pyridyl-azo)-2-naphthol (PAN). Reagent for cadmium, cobalt, indium, nickel, vanadium and zinc. 0·5 g in 100 ml of ethanol.

4-(2-Pyridyl-azo)-resorcinol (PAR). Reagent for cobalt, calcium, bismuth, cadmium, mercury, nickel, vanadium and zinc. 0·5 g in 100 ml of ethanol.

Pyrocatechol. Reagent for titanium. 10 g in 100 ml of water freshly prepared.

Pyrogallol. Reagent for bismuth. 10 g in 100 ml of water freshly prepared.

Pyrrole. Reagent for selenite. 1 g in 100 ml of ethanol free from aldehyde.

Quinaldinic acid. Reagent for cadmium and zinc. Neutralize 1 g with M sodium hydroxide and dilute to 100 ml with water.

Quinalizarin. Reagent for magnesium. 0·05 g in 100 ml of 0·1 M sodium hydroxide soln.

Quinalizarin-sulphonic acid. Reagent for bismuth. 0·5 g in 100 ml of water.

Quinoline. Reagent for silicon. Dissolve 2 g in 80 ml of hot water, add 2·5 ml of conc hydrochloric acid, cool and filter, preferably through a paper pulp layer.

Resorcinol. Reagent for platinum. 1·0 g in 100 ml of water.

Rhodamine B. Reagent for antimony, gold and tungsten. 0·01 g in 100 ml of water.

Salicylaldoxime. Reagent for bismuth, iron, lead, nickel and zinc. 1 g in 5 ml of cold ethanol added in drops to 95 ml of warm water.

Saltzmann's reagent (Reagent for nitrite). 0·02 g of N-(1-naphthyl) ethylene-diamine hydrochloride and 5 g of sulphanilic acid in water containing 140 ml of glacial acetic acid. Make up to 1 litre with water.

Silver periodate. Reagent for acetate. Add to 25 ml of potassium periodate soln (2 g per 100 ml of water) 2 ml of silver nitrate soln (10 g per 100 ml) and 2 ml of conc nitric acid. Filter.

Sodium azide-iodine reagent. For sulphide, thiosulphate and thiocyanate. Dissolve 3 g of sodium azide in 100 ml of 0·05 M iodine soln.

Sodium carbonate-phenolphthalein reagent. For bicarbonate. 20 ml of phenol phthalein soln (see indicators) and 10 ml of 0·05 M sodium carbonate soln mixed and diluted to 100 ml.

Sodium diethyl di-thiocarbamate. Reagent for cadmium,

chromium, copper, lead, mercury, cobalt, manganese, nickel, uranium and zinc. 0·1 g in 100 ml of water.

Sodium dihydroxytartrate osazone. Reagent for calcium, 1 g in 100 ml of water.

Sodium rhodizonate. Reagent for barium, lead and strontium. 0·5 g in 100 ml of water.

Sodium stannite. Reagent for bismuth. Add before use equal volumes of stannous chloride soln and of a soln of 27 g of sodium hydroxide in 100 ml of water.

Sodium tetraphenyl-boron. Reagent for ammonium, caesium, potassium and rubidium. 0·5 g in 100 ml of water.

Sodium tungstate. Reagent for vanadate. 10 g in 100 ml of water.

Solochrome black (Eriochrome Black T). Reagent for calcium, cadmium, lead, magnesium, manganese and zinc. 1 g in 25 ml of ethanol and 75 ml of triethanolamine.

Sulphanilic acid-1-naphthylamine reagent. Reagent for nitrites. (*a*) 1 g of sulphanilic acid in 100 ml of 5 M ('concentrated') acetic acid. (*b*) 0·3 g of naphthylamine boiled in 100 ml of 5 M acetic acid.

Sulphosalicylic acid. Reagent for bismuth and iron. 25 g in 100 ml of water.

Tannic acid solution. Reagent for aluminium, beryllium, germanium and tin. 1 g in 100 ml of water.

Tetramethyl-diamino-diphenyl-methane (Tetra-base). Reagent for lead, manganese and iodide. 0·5 g in 20 ml of glacial acetic acid made up to 100 ml with ethanol.

Thioacetamide. Reagent for antimony, arsenic, bismuth, cadmium cobalt, lead, nickel and zinc. 7·5 g in 100 ml of water.

Thiourea. Reagent for antimony, bismuth, cadmium and tin. 10 g in 100 ml of water.

Titan yellow. Reagent for magnesium. 0·1 g in 100 ml of water.

o-Tolidene. Reagent for chlorine, gold and manganese. 0·1 g in 100 ml of 3 M hydrochloric acid.

Toluene-3:4-dithiol. See Dithiol.

7

Reagents used in Organic Chemistry

Acetic acid-sodium acetate reagent. Add a sat soln of sodium acetate in water to an equal volume of glacial acetic acid.

Ammoniacal silver nitrate. See Tollen's reagent.

Aqueous-alchoholic silver nitrate. Dissolve 4 g in 10 ml of water and add 90 ml of ethanol.

Bang's reagent. Estimation of glucose. 100 g of potassium carbonate, 66 g of potassium chloride and 160 g of potassium bicarbonate dissolved in succession in approx 800 ml of water. Add 4·4 g of cupric sulphate crystals and dilute to 1 litre. After 24 hours dilute 300 ml of this solution to 1 litre with sat potassium chloride soln, shake gently and use 24 hours later. Avoid the entry of air. 50 ml = 10 mg of glucose.

Barfoed's reagent. Test for glucose. 13·3 g of cupric acetate and 2 ml of glacial acetic acid in 200 ml of water.

Bartrand's reagent. Estimation of glucose. Four separate solns, each of 1 litre, are made by dissolving the following: (*a*) 200 g of Rochelle salt + 150 g of sodium hydroxide; (*b*) 40 g of copper sulphate; (*c*) 50 g of ferric sulphate + 200 ml of conc sulphuric acid; (*d*) 5 g of potassium permanganate.

Bial's reagent. Test for pentose. See Chapter 9.

Biuret test. For urea and proteins. Dissolve 0·75 g of copper sulphate in 1 litre of 2 M potassium hydroxide soln.

Brucke's reagent. Test for proteins. Saturate a soln of 50 g of potassium iodide in approx 200 ml of water with mercuric iodide, (approx 120 g) and dilute to 1 litre.

Cuprous chloride, acid; cuprous chloride, alkaline. See Bench reagents.

Cuprous chloride, ammoniacal. Test for acetylene. Dissolve 0·5 g of copper sulphate crystals in about 5 ml of 2 M ammonia soln.

Add drops of a soln of hydroxylamine hydrochloride until the deep blue colour changes to colourless.

Denige's reagent. Test for Citrates. See Chapter 9.

Diazo solution (benzene diazonium chloride). Test for salicylic acid. Dissolve 0·5 g of aniline in 1·5 ml of conc hydrochloric acid. Cool this in ice and add a soln of 0·5 g of sodium nitrite in 2 ml of water until the soln gives a blue colour with a starch-potassium iodide paper.

Dichromate soln. For oxidation. Add 250 ml of conc sulphuric acid to 750 ml of water and dissolve 100 g of sodium dichromate in it.

Dimedone. Test for aldehydes. 0·3 g in 50 ml of water and 50 ml of ethanol.

2:4-Dinitro-chlorobenzene. Test for mercaptans. Dissolve 20 g in 100 ml of hot ethanol.

Diphenylamine. Test for rayon. 0·2 g in 100 ml of conc sulphuric acid.

Esbach's reagent. Test for albumin. 1 g of picric acid and 2 g of citric acid in 100 ml of water.

Fehling's solution. Test for and determination of reducing sugars and aldehydes.

Solution A. 34·64 g of copper sulphate crystals in 500 ml of water.

Solution B. Quantities vary from formula to formula but the B.P. amounts are 176 g of sodium potassium tartrate and 77 g of sodium hydroxide in 500 ml of water.

Equal volumes of the two solutions, which are kept separate, are mixed when ready for use. 5 ml of each = 10 ml of soln ≡ 0·05 g of glucose.

Fenton's reagent. Test for tartrates. Sat ferrous sulphate soln, 10 volume hydrogen peroxide and excess of sodium hydroxide soln added in that order.

Ferric chloride, neutral. Test for salts of organic acid. Add drops of ammonia soln or sodium hydroxide soln to 0·5 M ferric chloride soln (see Bench solutions) until a faint permanent ppt forms and filter the solution.

Fischer reagent. For water in organic solvents. Dissolve 400 g of pure iodine, dried over conc sulphuric acid in 3150 ml of anhydrous methanol, add 1260 g of pure, dry pyridine and pass in sulphur dioxide until the weight increases by 320 g.

Folin's solution. For uric acid. Dissolve 500 g of ammonium sulphate, 5 g of uranyl acetate and 6 ml of glacial acetic acid in approx 800 ml of water and dilute to 1 litre.

Formaldehyde-sulphuric acid. See Marquis's reagent.

Fuchsin-sulphurous acid. See Schiff's reagent.

Glyoxylic acid. Test for proteins. Cover 10 g of magnesium powder with water, add slowly 250 ml of a sat soln of oxalic acid, cooling the mixture during the addition. Filter the soln, slightly acidify it with acetic acid and make up to 1 litre.

Gunsberg's reagent. Test for hydrochloric acid in gastric juices. See Chapter 9.

Hager's reagent. For alkaloids. Sat soln of picric acid in water.

Hanus's solution. For iodine number. Dissolve 13·2 g of pure iodine in 1 litre of glacial acetic acid, heating if necessary. To the cold soln add equivalent quantity of bromine.

Hydroxylamine hydrochloride. Test for esters. Sat soln in methanol.

Hydroxylamine reagent. Dissolve 35 g of pure hydroxylamine hydrochloride in 160 ml of water and make up to 1 litre with ethanol. It is 0·05 M.

Iodo-potassium iodide. Test for alcohols, and aldehydes. Dissolve 6 g of potassium iodide and 2 g of iodine in 100 ml of water.

Marme's reagent. Test for alkaloids. Dissolve 33 g of potassium iodide and 16 g cadmium iodide in 50 ml of water and make up to 100 ml with sat potassium iodide soln.

Marquis's reagent. Test for alkaloids. 10 ml of formaldehyde soln (formalin) in 50 ml of conc sulphuric acid.

Mayer's reagent. Test for alkaloids. Dissolve 1·36 g of mercuric chloride in approx 50 ml of water, add a soln of 5 g of potassium iodide in water and make up to 100 ml.

Millon's reagent. Test for protein. See Chapter 9.

Molisch's reagent. Test for carbohydrates. See Chapter 9.

Nickel oxide, ammoniacal. Reagent for silk. Precipitate nickel hydroxide by adding 2 M sodium hydroxide soln to a soln of 5 g of nickel sulphate in approx 100 ml of water, filter and wash the ppt and dissolve in mixture of 25 ml of 0·88 ammonia and 25 ml of water.

Ninhydrin. Test for amino acids. 0·1 g of indane-trione hydrate in 100 ml of water.

Nylander's reagent. For carbohydrates and estimation of glucose. Dissolve 4 g of Rochelle salt (sodium potassium tartrate) and 2 g of bismuth subnitrate in 100 ml of 2 M sodium hydroxide soln.

Obermeyer's reagent. Reagent for indoxyl. 4 g of ferric chloride in 1 litre of conc hydrochloric acid.

Pavey's solution. Reagent for glucose. Add 300 ml of 0·88 ammonia to a mixture of 60 ml of each of Fehling Solution A and B. Dilute to 1 litre.

o-Phenylene-diamine. Test for quinones and diketones. Dissolve 5 g in 100 ml of hot ethanol.

Phloroglucinol. Reagent for pentosans. 3 g in 100 ml of ethanol.

Phosphomolybdic acid. Reagent for alkaloids. Dissolve ammonium molybdate in conc nitric acid and add phosphoric acid. Dilute, filter, wash the ppt and boil with aqua regia. Evaporate to dryness and dissolve the residue in 2 M nitric acid.

Picric acid. See Chapter 9.

Schiff's reagent. Test for aldehydes. Dissolve 0·5 g of fuchsin (magenta) in 500 ml of water and decolourize it by passing sulphur dioxide through the soln. Alternatively add to the fuchsin soln 9 g of sodium bisulphite and 20 ml of 2 M hydrochloric acid.

Schulze's reagent. Test for cellulose. Make a concentrated soln of zinc chloride, potassium iodide and iodine in water.

Schweitzer's reagent. Reagent for dissolving cellulose. Boil a solution of 5 g of copper sulphate crystals in approx 100 ml of water and add slowly 2 M sodium hydroxide soln until precipitation is complete. Filter and thoroughly wash the ppt and dissolve in the minimum volume of 4 M ammonium hydroxide.

Seliwanoff's reagent. Test for fructose. 0·5 g of resorcinol in 1 litre of 3 N hydrochloric acid.

Sodium bisulphite. Reagent for aldehydes and ketones. Dissolve 380 g of sodium metabisulphite and make up to 1 litre.

Sodium nitroprusside. Test for aldehydes and ketones. 1 g in 100 ml of water, freshly prepared.

Stoke's reagent. 3 g of ferrous sulphate and 2 g of tartaric acid in 100 ml of water and add, before use, 0·88 ammonia soln until the ppt first formed is redissolved.

Sulphomolybdic acid. Reagent for glucosides. Dissolve 10 g of molybdic acid or sodium molybdate in 100 ml of conc sulphuric acid.

Tannic acid. Reagent for albumin and gelatin. Dissolve 10 g in 10 ml of ethanol and dilute to 100 ml with water.

Tollen's reagent. Test for aldehydes and reducing sugars. Add to 5 ml of 0·2 M silver nitrate soln two or three drops of 2 M sodium hydroxide soln and add 2 M ammonia soln until the ppt just dissolves.

Topfer's reagent. For determining hydrochloric acid in gastric juice. See Chapter 9.

Uffelmann's reagent. Test for lactic acid. See Chapter 9.

Wij's solution. For determination of iodine number. Dissolve 13 g of iodine in 1 litre of glacial acetic acid. Remove 25 ml of the solution and pass dry chlorine into the remainder until the deep red colour changes to orange. Add the 25 ml of original soln to remove excess chlorine.

8

Biochemical Solutions and Reagents

This chapter includes a wide variety of solutions and reagents used in biology for qualitative and quantitative procedures. The majority of bench solutions are common to biology and chemistry and are generally of the same concentration. To avoid duplication, therefore, cross-reference is given to these in Chapter 4. Mention is made of any particular biological uses. The arrangement is alphabetical.

Acetic acid. See Chapter 4. For a 3% soln (0·5 M) make up 27 ml glacial acetic acid to 1 litre with water.

Acetone. Fat solvent and dehydrating agent. See also Cleaning Solutions.

Alcohol, ethyl. See Ethanol.

Alkaline iodide. Winkler method for determining dissolved oxygen. Dissolve in water 500 g sodium hydroxide (or 700 g potassium hydroxide) and 135 g sodium iodide (or 150 g potassium iodide) and make up to 1 litre.

1:2:4-Aminonaphthol sulphonic acid. Reducing agent for determination of plasma inorganic phosphate. Dissolve 0·5 g 1:2:4:-acid in 195 ml 15% sodium bisulphite and add 5 ml 20% sodium sulphite. Soln keeps for only 2 weeks.

Ammonia. See Chapter 4.

Ammoniacal copper oxide. Schweitzer's reagent. See Chapter 7.

Ammonium carbonate. Sat soln.

Ammonium chloride. Sat soln.

Ammonium molybdate. Detection and determination of phosphate. See Chapter 4.

Ammonium oxalate. Precipitation of calcium in blood. Determination of calcium. Sat soln.

Ammonium sulphate. Precipitation of proteoses, protein (casein) in milk. Colorimetric determination of plasma protein. Sat soln 56 g 100 ml soln.

Ammonium thiocyanate. Determination of chloride in plasma and urine. 13 g/litre water.

Arsenomolybdate reagent (Nelson). Colorimetric determination of blood glucose. Dissolve 25 g ammonium molybdate in 450 ml water. Add 21 ml conc sulphuric acid, then 3 g sodium arsenate ($Na_2HAsO_4.7H_2O$) dissolved in 25 ml water. Mix, then allow to stand at 37°C for from 1 to 2 days. Store in a brown, glass-stoppered bottle.

Barfoed's solution. Test for reducing sugars. See Chapter 7.

Barium chloride. See Chapter 4.

Baryta water. Carbon dioxide absorbent (e.g. respiration experiments). See Chapters 3 and 4.

Benedict Osterberg reagent. Red colour with reducing sugars. Dry 36 g picric acid at 60°C and add to 500 ml 10% (2·5 M) sodium hydroxide in a 1 litre flask. Mix thoroughly, add 400 ml hot water. Make up to 1 litre when dissolved.

Benedict's solution. Detection of reducing sugars. Dissolve in water 17·3 g copper sulphate, 173 g sodium citrate, 100 g anhydrous sodium carbonate and make up to 1 litre with water. For quantitative sugar determinations make up as follows. Dissolve 18·00 g crystallized copper sulphate in approx 100 ml water. To within 1 g weigh out 200 g crystallized sodium carbonate (or 100 g anhydrous sodium carbonate), 200 g sodium or potassium citrate, and 125 g potassium thiocyanate and dissolve in water to give 800 ml soln, warming if necessary. Place this soln in a graduated litre flask and add copper sulphate soln rinsing flask twice, adding washings. Then add 6 ml 0·1 M potassium ferrocyanide (see Chapter 4) making up bulk to 1 litre at room temperature (25 ml of this soln is reduced by 50 mg glucose).

Benzidene. Test for blood. Sat soln in glacial acetic acid.

Bial's reagent (Sumner). Specific test for pentoses and glycuronic acid. Dissolve 6 g resorcinol in 200 ml 95% ethanol which contains 40 drops 10% ferric chloride (see Chapter 4).

Bleach. See calcium hypochlorite, Chapter 4.

Bromine water. Knoop's test for histidine. Sat soln. See Chapter 4.

Calcium chloride. To clot blood to which anticoagulant has been added and thus to show that calcium ions are a necessary part of the blood-clotting mechanism. 0·5 M (2·5%) used for colorimetric determination of plasma protein. Detection of fats. M contains 111 g/litre.

Calcium hydroxide. See Chapter 4.

Calcium hypochlorite. Sat soln. See Chapter 4.

Chloral hydrate–iodine. Test for starch. Add iodine crystals to soln of 40 ml chloral hydrate in 25 ml distilled water (or other quantities in proportion).

Chlorine water. Sat soln. See Chapter 4.

Chloroiodide of zinc. See Schulze's solution, Chapter 10.

Cobalt chloride. Cobalt chloride paper for moisture detection (e.g. demonstration of transpiration through stomata). 5% soln.

Collodion. Making collodion bags for dialysis and demonstration of osmotic phenomena. Dissolve collodion in mixture of equal parts 100% ethanol and ether.

Copper sulphate. $CuSO_4.5H_2O$. Bench strength 1% soln and variety of other strengths. Use very dilute soln (barely coloured) for Biuret test. See also Harding's copper sulphate reagent, 7 g $CuSO_4.5H_2O$/100 ml distilled water; used for blood glucose determination colorimetrically.

Copper tungstate. Solution used for precipitation of protein in blood for preparation of protein-free blood filtrate (see Somoygi, *J. Biol. Chem.*, 1930, **86,** 655; 1931, **90,** 725).

Cross and Bevan's reagent. Test for cellulose. 2 parts conc hydrochloric acid to 1 part zinc chloride crystals by weight.

Denigè's reagent. Test for citrate (in milk, for example). Dissolve with the aid of heat 50 g red or yellow mercuric oxide in mixture of 200 ml conc sulphuric acid and 1 litre water.

Diacetyl monoxine. 3% aq soln. Fearon's test for citrulline.

Diazo-uracil. Formed from amino-uracil during specific colour test for sucrose by adding 1 ml 1% sodium nitrite soln to 5 ml amino-uracil (0·2% 5:amino-uracil in N/20 hydrochloric acid). N.B. Sugars such as glucose, fructose, maltose, give only yellow-brown colour.

2:6 Dichlorphenolindophenol. In a 50 ml flask dissolve one tablet of the dye in a small quantity of distilled water and make up to 50 ml mark (1 ml soln is equivalent to 0·02 mg ascorbic acid). Used for determination of ascorbic acid in urine. For blood determination 10 ml of the soln is diluted to 100 ml with distilled water (1 ml soln is equivalent to 0·002 mg ascorbic acid.)

p-Dimethyl-aminobenzaldehyde. Test for indole and skatole. 5% soln in ethanol.

2:4 Dinitrophenylhydrazine. With a little dilute alkali gives a blue colour with pyruvic acid.

Diphenylamine. Test for fructose—intense blue colour. 20% soln in 95% ethanol. 0·1 g dissolved in 10 ml nitrogen-free sulphuric acid used for Molisch's test for nitrates. Diphenylbenzidene may also be used.

Disodium phenyl phosphate. Estimation of alkaline phosphatase in blood. Dissolve 1·09 g in water and make up to 1 litre.

Ehrlich's reagent. Colorimetric determination of urobilinogen in urine. Dissolve 0·7 g p-dimethylaminobenzaldehyde in 150 ml conc hydrochloric acid and 100 ml water.

Ether. Used as fat solvent and narcotizing reagent.

Fehling's solution. Test for reducing sugars. Will also give Biuret (see Chapter 7).

Ferric alum. Indicator for chloride determination in urine. Sat soln.

Ferric chloride. Detection of lactic acid in muscle, and acetone bodies in urine. Bench strength (see Chapter 4).

Ferrous sulphate. Colorimetric determination of urobilinogen in urine. Dissolve 20 g ferrous sulphate in 100 ml water (must be freshly prepared).

Folin-Wu alkaline copper solution. Folin-Wu colorimetric determination of blood glucose. Dissolve 40 g pure anydrous sodium carbonate in 400 ml water, add 7·5 g tartaric acid, shaking to dissolve. Then add 4·5 g pure crystallized copper sulphate. Allow to dissolve and make up to 1 litre.

Fouchet's reagent. Fouchet's test for bile in urine. Dissolve 25 g trichloroacetic acid in 100 ml water, add 10 ml 10% ferric chloride.

Glyoxylic acid. Test for tryptophan and tryptophan radicals. Reduce a sat soln of oxalic acid with magnesium powder. For details see Benedict, *J. Biol. Chem.*, 1909, **6**, 51.

Guaiac solution. Test for blood in extreme dilutions. Dissolve 0·5 g guaiac resin in 30 ml 95% ethanol.

Gunzberg's reagent. Indication of 'free hydrochloric acid' in gastric contents. Dissolve 2 g phloroglucinol and 1 g vanillin in 100 ml 95% ethanol.

Harding's copper sulphate reagent. Determination of blood glucose.

Two Solutions. A. Dissolve 13 g copper sulphate ($CuSO_4.5H_2O$)

in water and make up to 1 litre. B. Dissolve 50 g pure sodium bicarbonate in 700 ml water, add 40 g anhydrous sodium carbonate, stirring continuously. When solution is complete add a warm (60°C) soln of potassium oxalate (36·8 g dissolved in 120 ml water) and add 24 g potassium sodium tartrate dissolved in water. Wash this into a 1 litre flask and make up to 1 litre. Mix equal volumes of A and B when required.

Haupt solution. To make leaf skeletons. Boil soln of 150 g sodium carbonate in 500 ml water and add 70 g calcium hydrate, boiling for a further 15 minutes. Cool. Filter. Leaves are boiled in the solution for 1 hour.

Hessler's fluid. Coloured fruit preservative. Dissolve 200 g zinc chloride in 4 litres of water, add 100 ml formalin and 100 ml glycerine. Decant off liquid if necessary.

Hubl's iodine solution. Test for unsaturated fats. Dissolve 2·5 g iodine and 3 g mercuric chloride in 100 ml 95% ethanol.

Hydrobromic acid. Bench strength 0·1 M.

Hydrochloric acid. See Chapter 4.

Iodine in potassium iodide. Test for starch. See Chapter 4.

Iodine–phosphoric acid (Mangin). Test for cellulose; violet. To a solution of 0·5 g potassium iodide in 25 ml conc phosphoric acid, add a few crystals of iodine.

Lactic acid. 0·9% (0·1 M). 10·5 ml syrupy 85% lactic acid/litre.

Lactic acid. 0·5% 6 ml syrupy 85% lactic acid/litre.

Lactic acid. 0·1% 1·2 ml 85% lactic acid/litre.

Lead acetate, neutral. Use with sodium hydroxide as a test for sulphur. See Chapter 4.

Lime water. Carbon dioxide absorbent. See Chapter 4.

Lindt reagent. Test for phloroglucinol, glucosidic tannins. 5 mg vanillin, 0·5 ml 100% ethanol, 3·0 ml conc hydrochloric acid, 0·5 ml water.

Locke's solution. Demonstration of tissue respiration. Methylene blue colour change indicates oxygen consumption. Dissolve 0·9 g sodium chloride, 0·042 g potassium chloride, 0·048 g calcium chloride, 0·02 g glucose, 0·01 g methylene blue in water and make up to 100 ml. Soln approximates in composition and osmotic pressure to the inorganic constitutents of blood.

MacLean's reagent. Reagent for lactic acid. Dissolve 5 g ferric chloride in 100 ml 0·2 M mercuric chloride soln (see Chapter 4). Add 1·5 ml conc hydrochloric acid.

Magnesium sulphate. Detection of fat—forms an insoluble soap. Sat soln.

Magnesium uranyl acetate. Ppt with cold sodium solutions. See Chapter 4.

Manganous sulphate. Winkler method for determining dissolved oxygen. Dissolve in water 480 g $MnSO_4.4H_2O$ (or alternatively 400 g $MnSO_4.2H_2O$) and make up to 1 litre.

Mercuric chloride. Precipitation of protein. 7 g 100 ml soln (0·25 M).

Mercuric sulphate. 10 g $HgSO_4$ dissolved in 100 ml 5% sulphuric acid.

Metaphosphoric acid. HPO_3. Protein precipitant. 0·6 M.

Methylamine hydrochloride. Test for maltose. 5% aq soln. Add 1 drop per ml of neutralized soln, boil for 30 sec and add approx 1 drop per ml of 20% sodium hydroxide soln. Yellow colour (warm) to carmine (cold).

Millon's reagent. Test for proteins and tryptophan. Dissolve 1 part by weight of mercury in 2 parts conc nitric acid. When solution complete add 2 parts water.

Molisch's reagent. General test for carbohydrates. Also test for arginine and arginine radicals in proteins using it with a few drops of sodium hypochlorite and sodium hydroxide. Dissolve 5 g α-naphthol in 100 ml ethanol.

Molybdic acid. Determination of plasma inorganic phosphate. *Two Solutions.* A. Dissolve 25 g ammonium molybdate in 200 ml water. Pour into a 1 litre flask containing 500 ml 5 M sulphuric acid with washings and make up to 1 litre with water. B. Repeat procedure for A but use only 300 ml acid.

α-Naphthol. See Molisch's reagent.

β-Naphthoquinone sodium monosulphate. Use with 10% sodium hydroxide and chloroform, test for indole in faeces. 2% soln.

Naphthoresorcinol. Specific test for glycuronic acid and glycuronates. 1% soln in ethanol.

α-Naphthylamine. See Chapter 6.

Ninhydrin. See Chapter 7.

Nitric acid. See Chapter 4.

p-Nitrobenzene-azo-resorcinol. See Chapter 6.

Obermayer's reagent. See Chapter 7.

Orcin. Test for inulin. 0·5% soln in 90% ethanol.

Osmic acid. Osmic acid test for unsaturated fats. 1% soln.

Oxalic acid. See Chapter 3.

Phenol reagent. Colorimetric determination of plasma protein. Dissolve 100 g sodium tungstate ($Na_2WO_4.2H_2O$), 25 g sodium molybdate ($Na_2MoO_4.2H_2O$) in 700 ml water. Add 50 ml phosphoric acid and 100 ml conc hydrochloric acid. Reflux for 8 hours, add a few drops of bromine if soln green and boil off excess. Filter if necessary.

Phenol solution. Estimation of blood alkaline phosphatase. Dissolve 1 g crystalline phenol in 0·1 M hydrochloric acid. Make up to 1 litre with more acid.

Phosphate solution (standard). Colorimetric determination of plasma inorganic phosphate. Dissolve 0·351 g potassium dihydrogen phosphate (pure) in approx 900 ml water. Add 10 ml 10 N sulphuric acid and make up to 1 litre with water. 0·4 mg phosphorus is contained in each 5 ml soln.

Phosphomolybdic acid. Folin-Wu determination of blood glucose. Place 35 g molybdic acid, 5 g sodium tungstate and 200 ml 10% sodium hydroxide in a flask with 200 ml water and boil the mixture for half an hour. Cool and dilute to 350 ml with distilled water before adding 125 ml 85% phosphoric acid. Make up to 500 ml.

Phosphotungstic acid. Dissolve in distilled water 50 g phosphotungstic acid and 30 ml conc sulphuric acid. Make up to 1 litre with distilled water.

Picric acid. Sat soln. Use: precipitation of protein. Determination of creatinine and creatine in urine. The picric acid must be pure. This may be achieved in the following way. Dissolve 50 g impure picric acid in 700 ml distilled water by heating and add 10 ml conc hydrochloric acid to liquid when boiling. After cooling decant off liquid and wash the crystals with 100 ml distilled water. Recrystallize, filter through a Buchner funnel, wash with 150 ml distilled water and dry with filter paper.

Picrolonic acid. See Chapter 6.

Potassium carbonate. Sat soln.

Potassium dichromate. $K_2Cr_2O_7$. 2·5% soln. Use with dilute sulphuric acid for determining acetaldehyde formation.

Potassium ferricyanide. Oxidation of haemoglobin to methaemoglobin. 0·1 M freshly prepared.

Potassium ferrocyanide. Indicator for phosphate in urine determination. 0·1 M soln.

Potassium mercuric iodide. See Nessler's reagent.

Potassium oxalate. Blood anticoagulant. Approx M.

Potassium permanganate. For determination of serum calcium use 0·001 M. See Chapter 4.

Potassium persulphate. $K_2S_2O_8$ 1% soln. Fearon's test for citrulline.

Pyrogallol (alkaline). Oxygen absorbent. Dissolve 28 g pyrogallol in 100 ml water. Separately dissolve 50 g potassium hydroxide in 100 ml water. Mix the two solutions prior to use.

Robert's reagent. Precipitation test for proteins. Mix conc nitric acid and a sat soln magnesium sulphate in the ratio 1:5.

Russow's reagent. Callus, deep brown. 1 part chlor-zinc iodide, 1 part iodine soln in potassium iodide (q.v).

Ruthenium sesquichloride (Ruthenium red). Pectin and pectic mucilage, deep red. Stain used after sections have been steeped in acid alcohol (1 part hydrochloric acid: 3 parts ethanol) for 24 hours and ammonia for a few hours. 1 in 5000 soln in water. Solution must be kept in the dark.

Salicylanilide ('Shirlan'). Antiseptic, prevents mould growth. 0·04% soln.

Schiff reagent. Test for lignin, aldehydes and cuticle. See Chapter 7.

Schimper's solution. To 100 ml distilled water add 160 g chloral hydrate, then a few drops of iodine to colour the soln.

Schweitzer's reagent (Ammoniacal copper oxide). See Chapter 7.

Seliwanoff's reagent. Specific test for fructose, though glucose will give a positive result on prolonged warming (see Chapter 7).

Silver nitrate. Bench strength is used for the detection of chloride in blood, etc. (see Chapter 4). For quantitative determination use 29·061 g dissolved in water and made up to 1 litre. 1 ml of this soln is equivalent to 0·01 g NaCl and 0·006 g Cl′. For the Whitehorn procedure dissolve 2·905 g pure silver nitrate in water and make up precisely to 1 litre. 1 ml is equivalent to 0·001 NaCl.

Sodium bisulphite. See 1:2:4:aminonaphthol sulphonic acid.

Sodium carbonate. A wide variety of strengths are employed. 0·50–0·1 M for the investigation of optimum activity ranges of digestive enzymes. 0·5 M as antidote to acid pipetted into mouth (see first aid appendix). M for Pauli's test for histidine. Sat soln for estimation of alkaline phosphatase in blood.

Sodium chloride. See Chapter 4.

Sodium citrate. Blood anticoagulant. Sat soln.

Sodium cobaltinitrite. See Chapter 4.

Sodium fluoride. Preservative (e.g. of intestinal mucosa). 2% soln.

Sodium hydroxide. 10 M is used for the saponification of fats and Biuret test for proteins. 0·5 M for blood creatinine determinations (see Chapter 4).

Sodium hypobromite. See Chapter 4.

Sodium nitrite. 0·13 M. See diazouracil test for sucrose.

Sodium nitroprusside. Used in test for creatinine and acetone in urine. Also test for – SH group in muscle tissue. Use with ammonium hydroxide and a little acetic acid. Blue colour with pyruvic acid (see Chapter 4).

Sodium phosphate. See Chapter 4.

Sodium picrate (pure). Determination of blood creatinine. 1% soln. Folin's method for preparing pure sodium picrate is as follows. Place 100 g moist picric acid in 500 ml flask, add 100 ml acetone and dissolve picric acid by immersing flask in warm water. Then add 4 g norite, shake and filter. Do not allow soln to evaporate. Dissolve in a 1 litre flask 50 g anhydrous sodium carbonate and 20 g sodium chloride, then add acetone solution of picric acid. When evolution of CO_2 ceases, cool the flask in cold water for half an hour, filter using a Buchner funnel. Wash ppt with 500 ml 7% sodium hydroxide, then transfer to flask and add 500 ml boiling water, 4 g anhydrous sodium carbonate, then slowly, stirring all the time, 30 g sodium chloride. When dissolved cool, filter, then wash with 7% sodium chloride, followed by 2% sodium chloride, then a little methanol. Dry the ppt at room temperature.

Sodium sulphate (isotonic). Determination of blood glucose. Dissolve 1·5 g anhydrous Na_2SO_4 in 100 ml water.

Sodium sulphite. See 1:2:4:amino sulphonic acid.

Sodium tungstate. Determination of blood glucose. $Na_{10}W_{12}O_{41}\cdot 28H_2O$. Dissolve 10 g in 100 ml water.

Starch solution. See Chapter 5.

Sucrose. M 342 g/litre—used for suction pressure experiments. 5–10% aq soln for pollen tube culture at a temperature of 20–21°C.

Sulphanilic acid. See Chapter 6.

Sulphosalicylic acid. Test for protein in urine—albumin, proteose, and Bence-Jones protein (see Chapter 6).

Sulphuric acid. See Chapter 4.

Takayama's solution. Haemochromogen test for blood. 3 ml

2·5 M sodium hydroxide, 3 ml pyridine, 3 ml sat glucose soln, in 7 ml water. Make up fresh.

Tannic acid. Dissolve in distilled water 100 g tannic acid, 25 g sodium acetate, 25 g sodium chloride, 50 ml glacial acetic acid and make up to 1 litre.

Tauber's reagent. Specific test for pentoses (e.g. arabinose) and pentose compounds (e.g. riboflavin, nucleic acid). Dissolve 1 g benzidene in 25 ml glacial acetic acid. Solution keeps for 4 days.

Thymol solution in ethanol. 5% soln.

Toluene. An excellent urine preservative.

Topfer's reagent. Test for 'free hydrochloric acid' in stomach extract. Dissolve 0·5 g dimethylaminoazobenzene in 100 ml 95% ethanol.

Trichloracetic acid. Colorimetric determination of plasma protein. 20% soln. Use 10% soln as protein precipitant.

Tungstic acid. Blood protein precipitant. Prepare from 10% sodium tungstate and 19 ml/litre strength sulphuric acid.

Uffelmann's reagent. Test for lactic acid. Add ferric chloride to 1% soln phenol until amethyst blue colour is produced. Add 5 ml soln under test to 5 ml reagent.

Uranium acetate. Determination of phosphate in urine. Dissolve 35·0 g uranium acetate ($UO_2(C_2H_3O_2)_2.2H_2O$) in 1 litre water.

Uric acid reagent (Folin). Add 750 ml water to 100 g sodium tungstate in a 2 litre flask. Shake to dissolve, then add 80 ml 80% syrupy phosphoric acid. With a funnel in the mouth of the flask boil gently for 2 hours and add a few drops of bromine if the fluid is dark in colour. Boil off bromine and make up accurately to 1 litre in a marked flask (see Chapter 7 for alternative recipe).

9

Solutions in Histology

This chapter includes a wide variety of histological solutions and reagents including fixatives, stains, dehydrating agents, clearing agents, macerating fluids, mountants, narcotizing reagents and preservatives. This can only be a representative selection for the number of histological receipts is legion, but every endeavour has been made to include the most commonly used ones.

Solutions should be looked after carefully in order to achieve consistently satisfactory results with them. Stains, in particular, are liable to deteriorate unless they are kept away from direct sunlight and at room temperature, avoiding damp or very warm situations. The growth of moulds in bottles of aqueous stains can be prevented by adding an excess of chloroform to each bottle. With care, therefore, the deterioration can be kept to a minimum. Although certain materials must be made up as required, some stains, such as methylene blue and those of the Romanowsky group, benefit by keeping, due to oxidation. Stains should be kept away from ammonia and acids. Slight changes of pH may affect the desired results of staining. Providing that the containers are kept tightly closed, dry stains will keep almost indefinitely. It is advisable to take special care when weighing out dry stains and subsequently making up the solutions. Clothing is easily ruined.

Acetic acid. 1 ml glacial acetic acid made up to 100 ml with distilled water. Use: an excellent fixative for nuclei and chromosomes.

Acetic aniline blue (Hoffmann's blue). Dissolve 1 g aniline blue ws in 1 ml glacial acetic acid and 98 ml 50% ethanol. Use: stain for protoplasm, sieve plate callose.

Acetic iodine green. Mix 1 ml glacial acetic acid with 99 ml iodine green (see also Iodine green).

Acetic lacmoid (La Cour). Place 1 g lacmoid and 55 ml distilled water in a flask, heating to dissolve, then add 45 ml acetic acid and cool.

Acetic methylene blue. Dissolve 0·2 g methylene blue in 99 ml distilled water and 1 ml glacial acetic acid. Use: for discharging coelenterate nematocysts.

Acetic methyl green. Add 1·5 ml glacial acetic acid to 98·5 ml distilled water followed by enough methyl green to produce a transparent, weak sea blue solution. Use: to kill and stain protozoa.

Acetic orcein. Place 1 g orcein and 45 ml glacial acetic acid in a flask and plug with cotton wool before bringing almost to the boil. After cooling add 55 ml distilled water. Use: connective tissue stain.

Aceto-carmine (Schneider). Place 0·5 g carmine and 55 ml distilled water in a 200 ml flask, bring to the boil, add 45 ml glacial acetic acid, plug flask with cotton wool, boil again, cool and filter. Use: serves both as stain and fixative. Particularly good for protozoa and nuclei.

Acetone-xylene. Mix 1 vol anhydrous acetone with 4 vols xylene. Use: rapid clearing following 90% ethanol.

Acid alcohol. Mix 100 ml 70% ethanol and 3 ml hydrochloric acid. Use: differentiating agent.

Acid alcohol decolourizer (Lempert). Mix 0·5 ml pure hydrochloric acid, 0·5 g sodium chloride, 75 ml pure methanol and 25 ml distilled water.

Acid fuchsin (Mallory). ½% aq acid fuchsin soln (replace distilled water by ethanol (50%) to give alc soln). Use: an excellent general stain. Counterstain.

Acid green. See light green.

Acid magenta. See acid fuchsin.

Acid water. Mix 0·5 ml conc hydrochloric acid with 100 ml distilled water.

Albert stain no. 1. Mix 0·15 g toluidine blue, 0·2 g malachite green, 1 ml glacial acetic acid, 2 ml 95% ethanol and 100 ml distilled water. Use: stain for *C. diphtheriae*.

Albert stain no. 2. Add 2 g iodine and 3 g potassium iodide to about 25 ml distilled water. When completely dissolved add 275 ml distilled water. Use: stain for *C. diphtheriae*.

Alcian blue. Dissolve 1 g alcian blue in 100 ml distilled water. Add a crystal of thymol or a few drops of chloroform to preserve. Use: clear, permanent stain for mucin.

Alcohol-xylene. Mix 50 ml ethanol with each 50 ml xylene. Use: clearing agent.

Algal mountant. Lactophenol with 1·0% copper chloride.

Algal preservative. 90 ml ethanol, 5 ml formalin, 2·5 ml glycerine, 2·5 ml glacial acetic acid, 1·0 g copper chloride, 1·5 g uranium acetate.

Alizarin red S. 0·1% solution in water. Use with 100 ml 2% potassium hydroxide, 100 ml 0·2% formalin, and 100 ml glycerine. Use: bone formation (for details see Hood and Neill, *Stain Tech.*, **23,** 209–18).

Alkaline alcohol. Add 0·5 ml ammonium hydroxide (sg 0·880) to 99·5 ml 90% ethanol.

Alum carmine. Mix 10 g carmine, 25 g ammoniacal alum and 500 ml distilled water. Bring to the boil, cool, filter and make up to 500 ml with distilled water. Add 5 ml formaldehyde. Use: stain for small animals other than marine ones.

Ammoniacal fuchsin. Add 5% alcoholic basic fuchsin to ammonium hydroxide (sg 0·0880) to produce a permanent pale yellow colour. Filter. Use: stain for lignified tissue.

Ammonium bichromate. 1–5% soln in water. Use: very useful for hardening nerve tissue.

Ammonium oxalate. 0·5% soln used after tissues have been steeped (macerated) in 1:3 mixture hydrochloric acid and 100% ethanol for 24 hours.

Amyl acetate. Clearing agent.

Aniline blue WS. The composition varies with the use: (*a*) 1% soln in 85% alcohol, use: stain for cellulose; (*b*) 0·5 g aniline blue ws in 100 ml distilled water, use: stain for algae and fungi; (*c*) Boil 2 g aniline blue ws in 100 ml distilled water, add 2 ml acetic acid, cool and filter, use: stain for collagen, after using 1% phosphomolybdic acid as mordant. Aniline blue acetic is also known as Aniline blue (Masson).

Aniline blue-orange (Mallory). 2·5 g aniline blue ws, 10 g orange G., 5 g phosphotungstic acid and 500 ml distilled water. Use: connective tissue stain. Stains collagen.

Aniline blue-orange G. acetic (for Azan stain). Dissolve 0·5 g aniline blue ws, 2 g orange G., in 100 ml 8% acetic acid by warming. Filter when cold. Use: connective tissue stain. Stains collagen.

Aniline gentian violet. Warm 175 ml distilled water to about 40°C and stir in 10 g gentian violet and 5 ml aniline. Add 20 ml 95% ethanol and filter. Use: plant tissues, mitotic figures. Gram stain for micro-organisms.

Aniline oil. Clearing agent. Not for use after osmic acid fixation.

Aniline red. See Basic fuchsin.

Aniline sulphate. To a saturated solution in distilled water, after filtering, add a few drops of sulphuric acid. Use: stain for lignified tissue—specific for lignin.

Aniline xylene. Mix equal parts.

Archibald's stain.

Two Solutions. A. 0·5 g thionin, 1·0 ml formalin, in 100 ml 2% phenol in distilled water. B. 0·5 g methylene blue, 1·0 ml formalin, in 100 ml 2% phenol in distilled water. Allow to stand for 24 hours and mix equal parts prior to use. Use: micro-organisms, especially plague preparations.

Auramine phenol (Lempert). Warm a 3% aq soln of phenol to about 30°C and add 0·3 g auramine. Shake well and filter. Use: fluorescent stain for micro-organisms, especially turbercle bacilli and trypanosomes.

Azan stain no. 2. See Aniline blue-orange G. acetic.

Azo black (Chlorazol black E). 0·5% azo black in water or 70% ethanol. Use: alcoholic soln most satisfactory for general histology and cytology. Aq soln recommended for chromosomes.

Azocarmine (for Heidenhain Azan stain). 0·5 g azocarmine B or G, 1 ml glacial acetic acid in distilled water; make up to 100 ml. Use: connective tissue stain.

Azure A–Erie Garnet. Mix rapidly 40 ml 1% aq azure A and 20 ml ½% aq Erie garnet B. Filter. Refilter each month. Use: frozen sections.

Baker's softening fluid. 10 ml glycerol, 54 ml 96% ethanol and 36 ml distilled water. Use: softening of animal structures.

Basic fuchsin. Dissolve 0·1 g basic fuchsin in 150 ml distilled water and add 1 ml 70% ethanol. Use: nuclear stain. Bacteria and fungi. Constituent of many other stains e.g. ammoniacal fuchsin (q.v.).

Bensley's fixative. Mix 4 parts 2·5% mercuric chloride in distilled water with 1 part 2% osmic acid. Use: suitable for mitochondria.

Benzene. Clearing agent.

Benzene-phenol. 5 g phenol in 100 ml benzene. Use: clearing agent. Less likely to cause milkiness than xylene.

Benzyl violet. See Methyl violet 6B.

Berlese's fluid (Gum chloral). Dissolve 15 g gum arabic in 20 ml distilled water. Add 10 ml glucose syrup, then up to 160 g chloral hydrate to saturation and 5 ml glacial acetic acid. Use: mountant for acarina and insects.

Best's carmine. Place 2·5 g carmine, 1·25 g potassium carbonate, 6·25 g potassium chloride and 75 ml distilled water in a beaker. Stir and boil gently for a few minutes. Cool and add 25 ml liq ammonia fortis. DO NOT FILTER. Keep in a refrigerator. Use: stain for glycogen.

Biebrich scarlet acetic. Dissolve 0·2 g Biebrich scarlet in 100 ml 1% acetic acid. Use: plasma stain. Connective tissue.

Biebrich scarlet (glycerin-ethanol). Mix 25 ml 1% aq. Biebrich scarlet, 10 ml glycerin, 40 ml 95% ethanol and 425 ml distilled water.

Bismarck brown Y. Dissolve 0·3 g Bismarck brown Y in 100 ml distilled water. (Water may be replaced by 95% ethanol.) Use: nuclear stain. Especially suitable for plant tissues. 1% soln for whole mounts of small organisms.

Borax carmine (Grenacher). Boil together 2 g carmine, 8 g borax, and 200 ml distilled water for 30 minutes. Cool and make up to 200 ml before adding 200 ml 70% ethanol. Use: good general stain for plant and animal tissue.

Borax methylene blue (Unna). Heat 500 ml distilled water to 60°C and stir in 5 g methylene blue and 5 g borax. Allow to cool slowly. Solution improves with age. Use: connective tissue stain, Negri bodies.

Borax methylene blue (Masson). Heat 500 ml distilled water to 60°C and stir in 10 g methylene blue and 25 g borax. Allow to cool slowly. Solution improves with age.

Boric acid (Goodrich). Sat soln in saline containing 2 drops Lugol's iodine (q.v.) per 25 ml soln. Use: macerating fluid.

Borrel's methylene blue.

Two Solutions. A. 1 g methylene blue in 100 ml distilled water. B. 0·5 g silver nitrate in 100 ml distilled water. Dissolve silver nitrate in a 500 ml flask and add slowly, stirring constantly, 3% sodium hydroxide until no more ppt is formed. Allow ppt to settle before decanting liquid. Wash ppt with distilled water until washings are no longer alkaline to litmus. Add solution A to washed silver hydroxide and boil for 5 minutes, cool and filter. Use: haematological stain with 5% aq eosin. Nerve tissue.

Bouin fixative (Picro-formol). 75 ml sat aq picric acid soln, 25 ml commercial formalin, and 5 ml glacial acetic acid. (There are a number of modifications—see also Duboscq-Brasil.) Use: an extremely valuable fixative, especially before staining with Heidenhain's iron haematoxylin.

Breed's stain (Methylene blue (Breed)). Dissolve 0·3 g methylene blue in 30 ml 95% ethanol and add 100 ml 2½% phenol in distilled water. Use: bacteria in milk smears.

Brilliant cresyl blue (alcoholic). 0·3% brilliant cresyl blue in 100% ethanol. Use: particularly useful for blood. Also as intra vitam and supra vitam stain.

Brilliant cresyl blue (**aqueous**). 1% brilliant cresyl blue in 0·9% sodium chloride.

Brun's glucose mountant. 40 g glucose, 10 ml glycerin, 10 ml 2% thymol in ethanol, and 140 ml distilled water.

Canada balsam. Resin dissolved in either xylene or benzene. Use: most frequently used medium for mounting.

Carbol fuchsin (Ziehl-Nielsen). Dissolve 1 g basic fuchsin in 10 ml 100% ethanol; dissolve 5 g phenol in 100 ml distilled water and mix this with the basic fuchsin solution. Use: bacterial stain, bacterial spores, and various cytoplasmic inclusions.

Carbol fuchsin dilute. 10 ml carbol fuchsin with 90 ml distilled water. Use: Gram stain.

Carbol gentian violet. Warm 500 ml distilled water to approx 40°C and stir in 5 g gentian violet (Micro) and 10 g phenol; add 10 ml 95% or 100% ethanol and filter. Use: Gram stain alternative to aniline gentian violet (q.v.).

Carbol pyronin methyl green. 0·04 g methyl green, 0·175 g pyronin, 2·5 ml 95% ethanol, 20 ml glycerin, and add ½% phenol to make up to 100 ml. Use: plasma cell stain.

Carbol methylene blue. Dissolve 1·5 g methylene blue, 5·0 g phenol in 10 ml 100% ethanol and 100 ml distilled water. Use: bacterial stain.

Carbol thionin. Grind up 1 g thionin in 50 ml 95% or 100% ethanol, and add 5 g phenol and 500 ml distilled water. Filter. Use: stain for bacteria and fungi.

Carbol xylene. 125 g phenol (detached crystals) in 375 ml xylol.

Carmalum. Gently boil a mixture of 10 g carmalum powder, 5 ml glacial acetic acid, and 200 ml distilled water for 1 hour, cool and filter. Solution will keep with addition of antiseptic—0·1 g salicylic acid, or 0·05 ml formol/100 ml. Use: cell contents, non-woody plant tissue.

Carmine. Constituent of a number of stains. See aceto-carmine, alum carmine, Best's carmine, borax carmine (Grenacher). For use as an injection mass: add 5 g carmine to 10 ml distilled water, and leave overnight. Add 100 ml strong ammonia, shaking to dissolve the dye, then filter twice through glass wool, keeping for several days at 37°C until all trace of ammonia smell has gone. Filter again.

Carnoy fluid. 10 ml glacial acetic acid, 30 ml chloroform, 60 ml 100% ethanol. Use: fixative.

Cedarwood oil. Clearing agent.

Celloidin. 2, 4, 6, 8, 10% solns. Use: embedding.

Cellulose acetate. 12% cellulose acetate in acetone. Use: to soften up wood for sectioning.

Champy (dichromate fixative). 3% aq potassium dichromate soln 7 parts, 1% aq chromic acid soln 7 parts, and 2% aq osmium tetroxide soln 4 parts. Use: fixative. Good for cytoplasmic structures, especially mitochondria.

Chloral hydrate. 64 g in 40 ml distilled water. Use: narcotizing reagent. For small marine creatures use 0·1% aq soln.

Chlorazol black E. See Azo black.

Chlorazol paper brown B. 3% aq soln. Use: plant tissues other than those of monocotyledons.

Chloroform. Clearing agent. Narcotizing reagent.

Chlor-zinc-iodine. See Schulze's soln.

Chromic acid (Priestley). 5% soln. Use: macerating fluid for plant tissues.

Chromic acid (Ranvier). 0·2% soln. Use: macerating fluid for animal tissues.

Chromo-acetic. There are several formulae depending on the plant tissue to be fixed. (1) (*Schaffner's modification*) 0·7 g chromic acid, 0·3 ml glacial acetic acid in 99·0 ml distilled water. (2) 1 g chromic acid, 1 ml glacial acetic acid in 98 ml distilled water. Details of others may be found in *Elementary Microtechnique*—H. Alan Peacock, *The Microtomist's Vade-mecum*—Bolles-Lee (ed.

Gatenby and Beams) and *Plant Science Formulae*—McLean and Cook.) Use: 1. Suitable fixative for algae. 2. General solution for plant tissue.

Clove oil. Clearing agent.

Cocaine. 0·5–1·0% or 30 ml 2% aq soln, 10 ml 96% ethanol and 60 ml distilled water. Use: narcotizing reageant.

Colophonium alcohol. Colophonium ½% in 100% ethanol. Use: differentiator.

Congo red. Dissolve 1 g Congo red in 100 ml distilled water to which a few drops of ammonium hydroxide have been added. Use: plant tissue stain, especially suitable for rusts. Useful as vital stain, if used dilute (0·5% soln).

Corallin red. Prepare a sat soln in 4% aq sodium carbonate soln. Add a little camphor as preservative. Use: stain for callose.

Cotton blue-lactophenol. Dissolve 1 g cotton blue in 100 ml lactophenol. For use dilute 5 ml of this solution to 100 ml with lactophenol. Use: temporary stain for fungi and algae.

Cotton blue-magenta-lactophenol. 4 ml cotton blue 1% in lactophenol, magenta 1% in lactophenol, and 50 ml lactophenol.

Crystal violet. ½% aq soln. Use: stain for bacteria (e.g. Gram's stain), amyloid, mitotic figures.

Cyanine. Dissolve 1 g cyanine in 100 ml 95% ethanol, add 100 ml distilled water. (Erythrosin may be used as a counterstain.) Use: stain for lignified tissue.

Cyanol. Dissolve 1 g xylene cyanol (or patent blue V) in 100 ml distilled water, add 10 g zinc powder, and 2 ml glacial acetic acid. Boil until colour disappears and filter freshly before use. Soln keeps for several weeks. Use: cyanol-peroxidase reaction for haemoglobin, erythrocytes, and haemoglobin containing renal casts and globules.

Cyanosine. See Phloxine.

Dahlia. 0·25–1% aq soln. Use: stain for amyloid, fresh animal tissues and plant nuclei.

Dammar. Pale amber soln in xylene. R.I. 1·5191. Less acid mountant than Canada balsam.

Delafield's haematoxylin. Dissolve in a water bath 8 g haematoxylin in 250 ml 100% ethanol and filter. Separately dissolve by warming 16 g ammonia alum in 800 ml distilled water. Mix this with haematoxylin soln and add 200 ml glycerine, mixing thoroughly. Leave exposed to daylight for at least 6 weeks to ripen or alternatively add a soln of 0·4 g potassium permanganate dissolved in 10 ml warm distilled water. Use: good general stain for non-woody plant tissue and animal tissues. Also whole animal and plant mounts.

Dianil red R. See Congo red.

3:4 Dihydroxyphenylalanine ('Dopa'). 0·1% aq soln. Use: Dopa reaction.

Dioxan (di-ethylene dioxide). Dehydrating and clearing agent. N.B. Vapours are toxic.

Distilled water. Essential to keep a good stock and to check it regularly with bromothymol blue. Store away from ammonia and acids.

Duboscq-Brasil (alcoholic Bouin). 150 ml 80% ethanol, 60 ml formol, 15 ml glacial acetic acid and 1 g picric acid. Use: fixative, highly suitable for fixing arthropods containing parasites.

Eau de Javelle. Dissolve 10 g potassium (or sodium) carbonate in 100 ml distilled water, add 5 g calcium hypochlorite in 100 ml distilled water and shake thoroughly. Allow the precipitate to settle and decant before use. Use: removal of soft tissues surrounding hard skeletal elements.

Eau de Labarraque. As for Eau de Javelle but substitute sodium hypochlorite for calcium hypochlorite. Use: as Eau de Javelle. Highly suitable for chitinous structures.

Ehrlich's acid haematoxylin. Dissolve in water bath 8 g haematoxylin in 400 ml 100% ethanol and filter. Separately dissolve by warming 8 g potash or ammonia alum in 400 ml distilled water. Mix this soln with the haematoxylin soln and add 400 ml glycerin and 40 ml glacial acetic acid, mixing thoroughly. Leave exposed to daylight for at least 6 weeks to ripen or alternatively add a soln of 0·4 g potassium permanganate dissolved in 10 ml distilled water. Use: excellent stain for general animal histology.

Ehrlich's triacid stain. Mix in the following order, using the same unwashed measure, 70 ml sat aq soln orange G., 35 ml 20% aq soln acid fuchsin, 75 ml distilled water, 75 ml 100% ethanol, 62 ml 10% aq soln methyl green, 50 ml 100% ethanol, and 50 ml glycerin. After adding methyl green shake well. DO NOT FILTER. Use: blood stain.

Elastin stain (Weigert). Place 0·75 dry Weigert's elastin stain and 5 g dry silver sand in a mortar, add 100 ml 95—100% ethanol and grind for around 5 minutes. Transfer soln and residue to a flask, plug with cotton wool and boil in a water bath for 20 minutes. Cool, filter, and add 2 ml pure hydrochloric acid. Use: stain for elastin.

Elastin stain (Sheridan). Prepare as for Weigert's elastin stain, substituting Sheridan's dry elastin stain for Weigert's. Use: stain for elastin.

Eosin. Dissolve 1 g eosin ws in 100 ml tap water. Add chloroform as preservative. Use: good cytoplasmic stain. Counterstain to Ehrlich's haematoxylin. For alcoholic soln substitute 75% ethanol for water.

Eosin methylene blue. See Jenner stain.

Erythrosin. 1 g erythrosin in 100 ml tap water. Add chloroform to preserve. Use: Cytoplasmic stain. For alcoholic soln substitute 90% ethanol for water.

Ethanol. Dehydrating agent—various strengths from 50% to 100%. 10% strength used as a narcotizing reagent for freshwater creatures. Where no strength is indicated ethanol of 95% upwards may be used. Industrial methylated spirit 74° o.p is suitable in Great Britain.

Ethanol (Ranvier). 30%. Use: macerating fluid.

Ether. Narcotizing reagent.

Ethyl phthalate. Good for temporary dry mounts (e.g. bacterial films Gram stained). R.I. 1·5.

Ethyl urethane. 0·3–1·0%. Use: narcotizing reagent.

Farrant's medium. Soln of Gum arabic and glycerine. R.I. 1·4200. Use: mountant with fat stains and for aqueous mounting purposes.

Fast green F.C.F. 0·5% in 95–100% ethanol. Alternatively 1 g fast green F.C.F. may be dissolved in 100 ml clove oil. Use: useful for non-lignified tissues. Feulgen reaction for chromatin containing thymonucleic acid.

Fast red 7B (Fettrot 7B) P.G. soln. Prepare sat soln as Sudan black (P.G. soln.). Use: fat stain.

Fat ponceau G. See Sudan III.

Feulgen solution (Schiff reagent). Boil 200 ml distilled water, add 1 g basic fuchsin. When dissolved cool to about 50°C and add 2 g potassium metabisulphite, dissolve and allow to cool to room temperature. Add 2 ml pure hydrochloric acid, plug flask with wool and leave overnight. Add ½ g decolourizing charcoal, shake and filter. The soln should be water white. Use: Feulgen reaction for D.N.A.

Feulgen sulphurous acid solution. 1 g potassium metabisulphite, 200 ml distilled water and 10 ml N hydrochloric acid. Use: Feulgen reaction for D.N.A.

Field's stain.

Two Solutions. A. Place in a flask in a warm water bath for half an hour 1·6 g methylene blue, 1 g azure 1, 2·6 g anhydrous disodium hydrogen phosphate, 2·6 g potassium dihydrogen phosphate and 1 litre distilled water. Filter. B. Dissolve in 1 litre of distilled water 2 g eosin, 2·6 g anhydrous disodium hydrogen phosphate, 2·6 g potassium dihydrogen phosphate. Filter. Use: rapid staining method for malarial parasites in thick blood films.

Flagella stain (Casares-Gil). Mix in a mortar 20 g tannin acid, 36 g hydrated alum chloride, 20 g zinc chloride and 3 g rosaniline hydrochloride. Add about 20 ml 100% ethanol. Mix and stir in a further 60 ml of ethanol to make 80 ml total.

Flagella stain (Loeffler). See Loeffler's flagella stain.

Flemming's fixative (strong). For 100 ml use 75 ml 1% aq chromic acid soln, 20 ml 2% aq osmium tetroxide soln, 5 ml glacial acetic acid (to be added just prior to use). **(weak)** 12·5 ml 1% aq chromic acid soln, 10 ml 2% aq osmium tetroxide soln, 0·1 ml glacial acetic acid in 78 ml distilled water.

Flemming's fixative (Taylor's formula). 0·2 ml 10% aq chromic acid, 2·0 ml glacial acetic acid, 1·5 ml 2% osmic acid in 2% chromic acid, 8·3 ml distilled water, and 0·15 g maltose (make up in small amounts when required). Use: specially suitable for root tip preparations and smears.

Fontana's stain.

Three Solutions. A. *Fixative* (*Ruge's solution*). 1 ml glacial acetic acid, 2 ml formalin, 100 ml distilled water. B. *Mordant.* 5% aq tannic acid soln, C. *Silver stain.* Dissolve 5 g silver nitrate in 100 ml distilled water.

Dilute 5 ml conc ammonia soln (liq ammon fort) with 45 ml distilled water; add 35 ml of the diluted ammonia to 90 ml of silver solution. Continue to add ammonia dropwise until the ppt formed is just redissolved. Then add sufficient of the remaining silver solution until the soln becomes slightly opalescent. Use: stain for spirochaetes.

Foot's silver solution. 12 ml 10% silver nitrate, 9 drops 10% sodium hydroxide. Add strong ammonia dropwise until ppt is just dissolved. Make up the volume to 42 ml with distilled water.

Formalin-dichromate. 50 ml 1% aq potassium dichromate, 50 ml 8% formaldehyde. Use: hardening of brain tissue.

Formol-acetic-alchol (F.A.A. fixative). 5 ml glacial acetic acid, 5 ml 40% formaldehyde, 90 ml 70% ethanol. Use: also preservative for algae.

Formol alcohol (Absolute-formol). 1 part formol, 9 parts 100% ethanol. Use: fixative.

Formol-calcium (Baker 1944). 1 g calcium chloride, 25 ml formaldehyde, 225 ml distilled water. Use: fixative.

Formol-saline. 8·5 g sodium chloride, 100 ml formalin, 900 ml distilled water. Use: fixative.

Formol sublimate fixative. 90 ml sat mercuric chloride soln, 10 ml formalin.

Fungi preservative. Dissolve in 1 litre water 1 g mercuric acetate and add 5 ml glacial acetic acid. Use: for preserving water insoluble pigment in fungi. Add 10 g neutral red acetate for soluble pigments.

Gallocyanin (Einarson). Dissolve 5 g chrome alum in a few ml warm water, add 0·15 g gallocyanin and boil for 10 minutes; cool, filter and make up to 100 ml with distilled water.

Giemsa stain. 3·8 g compounded Giemsa stain powder (or alternatively 3·0 g azur II eosin) and 0·8 g azur II are placed in a flask with 250 ml glycerin, and 250 ml distilled water. Plug neck loosely with cotton wool and warm in a water bath for 1 hour. Cool and filter. Use: haematological stain, for blood and blood parasites.

(N.B. Gurr's improved Giemsa stain, R. 66 should be obtained ready prepared.)

Gifford's softening fluid. 10 ml glacial acetic acid, 54 ml 95% ethanol, and 36 ml distilled water. Use: softening of plant material.

Gilson fixative. Mix 5 ml glacial acetic acid, 5 ml near 80% nitric acid, 100 ml 80% ethanol, 300 ml distilled water and 20 g dry zinc chloride.

Glycerine. 50 ml glycerine with 50 ml distilled water and 1 ml conc thymol soln. Use: mountant. Also soln of gelatine and glycerine in water—suitable for mounting fats.

Gossypimine. See Safranin aqueous.

Gothard's differentiator. 50 ml creosote, 40 ml cajaput oil, 50 ml xylene and 150 ml 100% ethanol.

Gram's iodine. Add 1 g iodine, 2 g potassium iodide to about 25 ml distilled water. When solution complete add further 275 ml distilled water to make a total amount of water of 300 ml. Use: Gram's stain for bacteria.

Gram's iodine (Kopeloff and Beerman). Dissolve 2 g iodine in 10 ml N/1 sodium hydroxide soln and add 90 ml distilled water.

Green algae preservative. Steep in 0·5–1% copper acetate in 2% formalin for 24 hours, followed by 5% formalin.

Gurr's N.R.G. stain. Obtain ready made.

Gurr's water mounting medium. Water-soluble gum in water. An aqueous mountant suitable for fat stains and aqueous mounting purposes. R.I. 1·4045.

Haemalum (Mayer). Dissolve in 1 litre of distilled water 0·2 g sodium iodate, 50 g alum, 50 g chloral hydrate, 1 g citric acid and add 10 ml. 10% haematoxylin in ethanol. Use: stain for animal tissue.

Harris haematoxylin. Dissolve 50 g ammonia alum in 500 ml distilled water by warming; add 25 ml freshly prepared 10% haematoxylin in ethanol. Bring to the boil and add 1·25 g mercuric oxide and stir in while boiling. Cool rapidly in a cold water bath. Filter. Mallory recommended the addition of 5% acetic acid. Use: good general histology stain.

Hayem's blood counting fluid (for red cells). Dissolve 0·3 g mercuric chloride in 120 ml warm distilled water; add 1·2 g sodium chloride, 3·0 g sodium sulphate, dissolve and filter.

Heidenhain's iron haematoxylin.

Two Solutions. A. 3 g iron alum in 100 ml freshly boiled water. B. 5 ml 10% alc haematoxylin in 95 ml distilled water. Solution A is a mordant. Use: good stain for plant and animal material.

Helly's fixative. 50 ml 3% aq potassium dichromate, 50 ml sat aq mercuric chloride soln, 1 g sodium sulphate. Immediately before use add 1 ml formaldehyde to each 20 ml.

Hiss's capsule stain.

Two Solutions. A. 5 ml fuchsin 10% in ethanol, with 95 ml distilled water. B. 20% aq copper sulphate soln.

Hitchcock-Ehrlich mixture.

Two Solutions. A. Dissolve 1 g malachite green in 50 ml distilled water. B. Dissolve 3 g acridine red in 150 ml distilled water. Mix one part of A with 3 parts of B just prior to use. Use: plasma cell stain.

Hoffmann's blue. See Acetic aniline blue.

Hofmann's ws violet. See Dahlia.

Hydrofluoric acid. 5% soln. Use: with plant material containing silica after fixation. Improves cutting.

Iodine. Add iodine to a sat aq soln of potassium iodide until saturated, then filter and dilute with distilled water to produce a pale golden brown soln. Use: temporary stain, especially useful for location of starch. Undiluted it is a useful fixative for simple algae.

Iodine green. Mix 1 g iodine green with 100 ml 70% ethanol. Use: with acid fuchsin counterstain for lignified tissues.

Iron aceto carmine (Belling). 54 ml aceto carmine, 2 ml ferric oxide sat soln in 45% acetic acid.

Iron alum. 2–3% aq soln in freshly boiled water. Use: mordant and differentiator for Heidenhain's iron haematoxylin.

Iso-propyl alcohol. Dehydrating agent.

Janus green B. 0·1 g (maximum) in 1 litre isotonic saline. Use: supravital stain especially suitable for mitochondria and blood.

Jeffery's macerating fluid. 50 ml 10% nitric acid, 10% chromic acid.

Jenner stain. Dissolve by stirring 0·5 g Jenner stain in 100 ml pure methanol. Use: haematological stain.

Kaiserling's fixative. 100 g potassium acetate, 200 ml glycerin, and 1 litre distilled water.

Kopeloff and Beerman formula for Gram's iodine. See Gram.

Lactic acid. 70–75% soln. Use: clearing agent.

Lactophenol. Dissolve 10 g phenol in 10 ml distilled water (not heating), and add 10 ml glycerine and 10 ml lactic acid. Use: mountant, especially for fungi (see also Cotton blue-lactophenol).

Lacto-fuchsin. 0·1 g acid fuchsin in 100 ml lactic acid. Use: for mounting fungi (superior to cotton blue-lactophenol, staining is more rapid).

Leishman stain. 0·15 g Leishman stain in 100 ml pure methanol. May be dissolved by: (1) place dye and methanol in a flask, plug neck with cotton wool and warm in a water bath for 15 minutes with occasional shaking; (2) triturate dye and methanol in a mortar for 15 minutes; (3) stirring cold for two hours. The solution improves as a granule stain with keeping. Use: haematological stain.

Lepine's fuchsin-safranin.
Two Solutions. A. 0·5 g basic fuchsin in 100 ml 50% ethanol. B. 0·2 g safranin in 100 ml water. Mix in two solutions. Use: stain for virus inclusions and elementary bodies.

Light green (Masson). 2 g light green, 2 ml glacial acetic acid, in 100 ml distilled water. Use: plant tissue stain. Also used in Masson trichrome stain.

Light green in clove oil. Heat in a water bath 1 g light green in 100 ml clove oil. Filter. Use: plant tissue stain. Most suitable for non-woody tissue.

Lissamine fast red B. (Picro-Lissamine) (Lendrum). 75 ml lissamine fast red B 1% aq with 25 ml 1% aq picric acid.

Loeffler's flagella mordant. Dissolve 20 g ferrous sulphate crystals in 40 ml distilled water by warming and add 100 ml 20% aq tannic acid soln and 10 ml basic fuchsin 10% in ethanol.

Loeffler's flagella stain. 20 ml basic fuchsin 10% in ethanol with 80 ml 3% aniline water. Use: stain for bacteria.

Loeffler's methylene blue. Warm 100 ml distilled water to about 50°C, stir in 0·5 g methylene blue and add 1 ml 1% potassium hydroxide soln and 30 ml ethanol (95–100%). Use: Bacterial stain.

Lugol's iodine. Add 1 g iodine and 2 g potassium iodide to 25 ml distilled water. When solution is complete add 75 ml of water to give total of 100 ml. Use: modification of Gram's method for bacteria.

Luxol fast blue—Cresyl fast violet (Kluver and Barrera).
Two Solutions. A. 1 g luxol fast blue in 1 litre 95% ethanol and 5 ml 10% acetic acid. Filter. B.1 g cresyl fast violet in 1 litre distilled water. Before using add 5 drops 10% acetic to every 30 ml and filter. Use: combined staining of cells and fibres in the nervous system.

MacConkey's capsule stain. 0·5 g dahlia, 1·5 g methyl green in 100 ml distilled water. Dissolve and add 10 ml fuchsin 10% alcoholic soln in 90 ml distilled water. Filter; stand two weeks before use. (Stain keeps for about 6 months.)

Magenta acid. See Acid fuchsin.

Magnesium chloride. Sat soln diluted with equal volume of sea water. Use: narcotizing reagent for marine animals.

Malachite Green. 1% aq soln. Use: stain for plant cytoplasm.

M'Faydean's methylene blue stain. A 'ripened' Loeffler's methylene blue. Expose Loeffler's methylene blue and air to light for several months until it acquires a purplish tinge. Use: stain for *B. anthracis*.

Mallory's aniline blue orange. See Aniline blue orange.

Mallory's phosphotungstic haematoxylin. Dissolve 1 g haematoxylin in a few ml distilled water, by heating and mix in 200 ml 10% phosphotungstic acid aq soln and remainder of 800 ml of distilled water. Ripen by standing for several weeks or artificially by adding 12·5 ml 1% potassium permanganate. Use: a good stain for many tissues, especially connective tissue.

Mallory-Heidenhain's (rapid one step stain). Add one at a time to 200 ml distilled water, stirring to dissolve, after each addition, 1 g phosphotungstic acid, 2 g orange G., 1 g aniline blue, 3 g acid fuchsin. Use: connective tissue stain.

Mann's stain. 35 ml methyl blue 1% aq soln, 45 ml 1% eosin, in 100 ml distilled water. Use: stain for acidophilic inclusion bodies.

Mann's differentiating alcohol. 100 ml 70% ethanol and 4 ml orange G. 7% aq.

Masson fuchsin. See Ponceau fuchsin.

Mayer's acid haemalum. See Haemalum—Mayer.

May-Grünwald stain. Dissolve by trituration 0·25 g May-Grünwald stain in 100 ml pure methanol. Use: blood and bone marrow smears.

Mercuric chloride. Sat aq soln. Use distilled water (isotonic saline is an alternative solvent). Use: fixative.

Merkel's fluid. 25 ml 1% chromic acid, 25 ml 1% platinic chloride, and 150 ml distilled water. Use: recommended alternative to chromo-acetic fixatives. Good for mitotic figures in higher plants.

Methyl benzoate. Clearing agent.

Methyl blue. 1% aq soln (use distilled water). Use: supravital stain for aquatic organisms. Counterstain for carmine, eosin.

Methylene blue. 1 g methylene blue in 100 ml distilled water, add 0·5 g sodium chloride. Use: stain for living organisms. For dead material use solution consisting of 0·3 g methylene blue dissolved in 30 ml 95% ethanol to which 100 ml distilled water has been added. Especially suitable for bacteria and delicate tissues. See also Borax methylene blue (Unna), Borrel's methylene blue, Loeffler's methylene blue, Neisser methylene blue, Newman stain.

Methylene blue-fuchsin (Gray).
Two Solutions. A. 50 ml 1% aq methylene blue with 50 ml pure methanol. B. 25 ml 1% aq basic fuchsin with 25 ml pure methanol. Dissolve dyes and mix solutions A and B. Use: bacteria in milk smears.

Methyl green. 1% alc soln (use 70% ethanol for plant tissues). Use: plant tissues, supravital stain for small organisms.

Methyl green-orange G. (Kardos stain). Dissolve in a warm water bath or by stirring 0·5 g Kardos mixture in 100 ml pure methanol. Use: stain for blood films.

Methyl green-pyronin (Jordan and Baker). Prepare a 0·5% aq soln methyl green and repeatedly extract with chloroform until chloroform extract is nearly colourless (at least eight extractions are necessary). *Buffer solution pH 4·8* 81 ml N/5 acetic acid with 119 ml M/5 sodium acetate. *Stain.* 37 ml 0·5% aq pyronin, 13 ml 0·5% methyl green (prepared as above) and 50 ml buffer soln. Solution keeps well for 4 months. Use: stain for R.N.A. and D.N.A.

Methyl green-pyronin (Pappenheim). 50 ml 1% aq methyl green, 50 ml 1% aq pyronin. Use: stain for *N. gonorrhoeae.*

Methyl violet 6B (Jensen). Dissolve 5 g methyl violet in 1 litre distilled water. Filter. Use: Jensen's modification of Gram's stain.

Methyl violet (Kopelhoff and Beerman). 30 ml 1% aq methyl violet 6B (or crystal violet) with 8 ml 5% sodium bicarbonate soln. Prepare freshly every 3 or 4 days.

Mucicarmine. 2·5 g dry mucicarmine in 100 ml 50% ethanol. Raise to boiling point in water bath, allow to boil for 3 minutes, cool and filter. Use: stain for mucin.

Muir's capsule stain mordant. 20 ml sat mercuric chloride soln (about 6%), 20 ml 20% tannic acid soln, 50 ml sat potash alum soln (about 10%).

Muller's fixative. 25 g potassium dichromate, 10 g sodium sulphate, in 1 litre distilled water.

Neisser Bismarck brown. Heat 500 ml distilled water to about 60°C, add 1 g Bismarck brown and dissolve. Use: stain for *C. diptheriae.*

Neisser chrysoidin. 1 g chrysoidin in 300 ml distilled water. Use: modified Neisser stain for *C. diptheriae.*

Neisser crystal violet. 1 g crystal violet, 10 ml ethanol (95–100%), 300 ml distilled water. Use: modified Neisser stain for *C. diptheriae.*

Neisser methylene blue. 10 ml 2% aq soln methylene blue, 4 ml ethanol (95–100%), 10 ml glacial acetic acid, 180 ml distilled water. Use: stain for *C. diptheriae.*

Neuroglia mordant (Anderson's).

Two Solutions. A. Dissolve 5 g sodium sulphite, 2·5 g oxalic acid, 5 g potassium iodide, and 2·5 g iodine in 100 ml distilled water. Add 5ml glacial acetic acid. B. 5 ml liq ferri perchlor fort in 90 ml distilled water. For use mix equal volumes of A and B.

Neutral red (Jensen). 1 g neutral red, 2 ml 1% acetic acid in 1 litre distilled water. Dissolve and filter. Use: Jensen's modification of Gram's stain for bacteria. 0·25% neutral red in neutral 100% ethanol is supravital stain for blood. 0·1 g dissolved in 1 litre of isotonic saline is a useful intra vitam stain.

Neutral red-fast green (Ollett). 90 ml 0·2% alc soln neutral red, 10 ml 0·2% soln fast green. Use: excellent bacterial stain.

Neutral red-light green (Twort). Grind in a mortar for 5 minutes, 1 g neutral red-light green about 10 g clean sand and 100 ml 60% iso-propyl alcohol. Filter. Use: stain for animal parasites, micro-organisms and tissues.

Newman stain (modified). Mix 54 ml 95% ethanol and 40 ml tetrachlorethane, warm in water bath to 55°C. Add 1 g methylene blue and shake to dissolve. Cool and add 6 ml glacial acetic acid. Add 8 ml 1% basic fuchsin in ethanol. Filter and keep well stoppered. Use: stain for bacteria in milk.

Nigrosine. 5% aq soln. Use: stain for bacterial spores and capsules.

Nile blue sulphate. Mix 1 g Nile blue, 0·5 ml pure sulphuric acid and 100 ml distilled water. Boil gently for 2 hours, occasionally making up the volume with water. Cool, filter and make up to 100 ml before use. Use: stain for fats and fatty acids. Dilute this is a useful supravital stain. Good for green algae.

Nitric acid. 20% (3 M). Use: macerating fluid.

Nitric acid/chromic acid (Jeffery). Equal volumes 10% nitric acid and 10% chromic acid. Use: macerating fluid for plant tissues.

Nitric acid in formol. 1–5% nitric acid in 5% formol. Use: decalcifying fluid.

Nuclear fast red (Kernechtrot). Dissolve 0·1 g nuclear fast red, 5 g aluminium sulphate in 100 ml distilled water by boiling. Use: nuclear counterstain (e.g. use with Sudan black).

Oil ponceau G. See Sudan III.

Oil red 4B. 0·5 g oil red 4B in 100 ml iso-propyl alcohol. Form sat soln by warming in water bath. Use: fat stain.

Oil red O. 0·5 g oil red O in 100 ml iso-propyl alcohol. Form sat soln by warming in water bath. Use: fat stain.

Opal blue. See Methyl blue.

Orange G. Recipe depends on use. For plant sections use 1% soln in distilled water or clove oil. As a cytoplasmic stain use 0·5% in distilled water or sat soln in 100% ethanol. 1% soln is a constituent of Flemming's triple stain.

Orcein (modified Unna). 1 g orcein (synthetic), 1 ml pure hydrochloric acid, with 100 ml 100% ethanol. Shake to dissolve, or stand overnight, filter. Use: stain for elastic fibres (see also Acetic orcein—la Cœur).

Osmium tetroxide. 1% aq soln. Use: fixative for animal tissue. Constitutent of many fixatives. Keep solution in the dark.

Pal's solution. 1 g oxalic acid, 1 g sodium sulphite, with 200 ml distilled water. Use: modified Anderson's method for neuroglia.

Pampel's fluid. 4 vols glacial acetic acid, 30 vols 40% formaldehyde, 6 vols 95% ethanol, 15 vols distilled water. Use: preservative, especially for insects (Imms).

Panchrome stain (Pappenheim). Place the following in a flask in a water bath and warm for 20 minutes—1 g methylene blue, 0·5 g toluidine blue, 1 g azur I, 0·5 g methylene violet, 0·75 g eosin, 250 ml pure methanol, 200 ml glycerin. Cool, add 50 ml acetone, shake well and filter. Use: blood stain.

Papanicolaou stain (EA 36). 11·25 ml 10% aq soln eosin, 2·25 ml 10% aq soln light green, 25 ml 1% alc soln Bismarck brown, 450 ml ethanol, 10 ml 10% aq soln phosphotungstic acid, and 0·3 ml sat aq lithium carbonate. Use: vaginal smears.

Papanicolaou OG6 stain. 32·5 ml sat aq soln orange G., 0·75 ml 10% aq soln phosphotungstic acid ethanol to 500 ml. Use: vaginal smears.

Pappenheim stain. See Methyl green pyromin—Pappenheim.

Pasini stain. 30 ml water blue-orcein, 30 ml eosin BA 2% in 50% ethanol, 4 ml 20% aq acid fuchsin, 25 ml glycerin. Use: connective tissue stain, particularly useful for embryology.

Phenol (Lendrum). 4% aq soln. Use: softening hard materials.

Phenylene brown. See Bismarck brown.

Phloroglucinol. 1 g in 100 ml distilled water. For lignified tissue add conc hydrochloric acid to alc soln until ppt forms. Solution is then ready for use.

Phloxine (Lendrum). 0·5 g phloxine, 0·5 g calcium chloride, in

100 ml distilled water. Use: general histological stain. Also demonstration of inclusion bodies.

Pianese stain. 0·5 g malachite green, 0·1 g acid fuchsin, 0·01 g Martius yellow, 50 ml 100% ethanol, 150 ml distilled water. Use: stain for fungi.

Picric acid. Sat aq soln for use as fixative (see also Bouin's fluid).

Picric alcohol (Masson). 60 ml sat alc soln picric acid, 30 ml 95% ethanol. Use: stain for chitinous structures.

Picro aniline blue. 0·1 g aniline blue ws in 100 ml sat aq picric acid. Use: plant tissue stain.

Picro-formol-alcohol ('Rossman's fluid'). 90 ml sat alc soln picric acid, 10 ml formaldeyde. Use: fixative.

Picro fuchsin. See Van Gieson.

Picro-Ponceau S. 10 ml 1% aq soln Ponceau S, 90 ml sat aq picric acid, 2 ml 2% acetic acid. Use: counterstain for Weigert's elastin stain and Ehrlich's acid haematoxylin.

Polychrome blue. Dissolve 1 g phenol in 10 ml 95% ethanol. Triturate with stain. Add approx 70 ml of 100 ml distilled water and stir. Pour into a flask adding washings from mortar using remainder of water. Allow to stand for 24 hours; filter. Use; stain for plant tissues.

Polychrome-methylene blue (Unna). Dissolve 1 g methylene blue in 100 ml distilled water, add 1 g potassium carbonate and 20 ml ethanol. Place in water bath and evaporate to 100 ml. Use: plasma cells, Nissl granules.

Polychrome toluidine blue (Martinotti). Warm 75 ml distilled water, stir in 0·5 g lithium carbonate, then 1·0 g toluidine blue, dissolve and add 20 ml glycerin and 5 ml 95% ethanol (previously mixed). Ripen for one month. Use: plasma cells, fungus hyphae in skin scrapings.

Polyvinyl lactophenol. Water miscible clearing and mounting medium. Small objects may be cleared directly from water.

Ponceau B. See Bilbrich scarlet.

Ponceau fuchsin (Masson). 0·3 g acid fuchsin, 0·7 g Ponceau de xylidine, 1 ml glacial acetic acid, in 100 ml distilled water. Use: Masson trichrome stain.

Ponder's stain. 0·1 g toluidine blue, 5 ml glacial acetic acid, 10 ml ethanol, add distilled water to 500 ml. Use: bacterial stain.

Potash alum. Sat aq soln. Use: constituent of Ehrlich's acid haematoxylin; may also be used as a diluent of this stain.

Potassium dichromate. Dissolve 1·5 g in 100 ml distilled water. Use: fixative for animal tissue (cytology), constituent of Champy, Helly, Regaud and Smith (all of which q.v.). N/10 soln contains 4·904 g/litre.

Potassium ferrocyanide. Make up a fresh mixture of 1·5% aq soln potassium ferrocyanide, and 0·5% aq hydrochloric acid using equal volumes of each. Use: location of iron compounds in sections.

Potassium hydroxide. 5% (M) aq soln. Use: saponification of fats, cleaning of skeletal preparations.

Potassium permanganate. 0·1% (0·006 M) aq soln. Use: fluorescent staining method for *Myco. tuberculosis* with auramine phenol stain and acid alcohol decolorizer (Lempert).

n-Propyl alcohol. Dehydrating agent.

Pugh's stain. 1 g toluidine blue, 25 ml glacial acetic acid, 10 ml ethanol, add distilled water to 500 ml. Use: stain for *C. diptheriae.*

Randolph's blood diluting fluid.

Two Solutions. A. 0·1 g methylene blue, 100 ml propylene glycol, 110 ml distilled water. B. 0·1 g phloxine, 100 ml propylene glycol. 100 ml distilled water. Immediately prior to use mix equal volumes. Mixture remains usable for up to 4 hours.

Regaud's fixative. 100 ml 3% potassium dichromate, 25 ml formaldehyde (neutralized by shaking with magnesium carbonate).

Regaud's haematoxylin. 1 g haematoxylin, 10 ml ethanol, 10 ml glycerin, 80 ml distilled water. Allow to ripen for 3 weeks, or prepare with 10 ml ripened 10% alc haematoxylin instead of the first two ingredients. Use: Masson trichrome stain.

Rongalite white(Leuco methylene blue). Add 5 drops pure hydrochloric acid to 100 ml ½% aq methylene blue. Warm and add 3 g rongalite. Cool and filter. Solution remains stable for a few days. Use: supravital stain for nerve cells, differential stain for normal and cancerous cells.

Rouselet's solution (Baker). 3 ml 2% cocaine hydrochloride, 1 ml 90% ethanol, 6 ml distilled water. Use: narcotizing reagent.

Saffron. Place 2 g saffron and 100 ml distilled water in a flask in a water bath and boil for 1 hour. Filter and add 1 ml formaldehyde and 1 ml 5% tannic acid. Use: connective tissue stain.

Safranin alcoholic. Dissolve 1 g safranin in 50 ml distilled water, add 50 ml ethanol. Filter. Use: stain for plant and animal tissue, elementary bodies, counterstain for Sudan black, constituent of Weigert elastin stain.

Safranin aqueous. Dissolve 1 g in 100 ml distilled water; filter. Use: bacterial spores, Negri bodies.

Sandiford's stain. 0·05 g malachite green, 0·05 g pyronin, 100 ml distilled water. Use: counterstain for gonococcus and meningococcus.

Scarlet R (alcoholic). 0·5 g scarlet R. in 100 ml 70% ethanol. Dissolve in a warm water bath and filter. Use: fat stain.

Scarlet R (Herxheimer). 1 g scarlet R. in 50 ml acetone and 50 ml 70% ethanol. Dissolve by shaking; filter. Use: fat stain.

Scarlet R (P.G. Soln). 1 g scarlet R. in 100 ml propylene glycol. Dissolve in a warm water bath; filter. Use: fat stain.

Schaudinn's fixative. 100 ml sat aq soln mercuric chloride with 50 ml 100% ethanol. Use: fixative for protozoa.

Schiff reagent. See Feulgen solution.

Schulze's macerating fluid. 1 g potassium chlorate in 50 ml conc nitric acid.

Schulze's solution. Mix 20 g zinc chloride with 8·5 ml distilled water. Dissolve by warming: cool, then add drop by drop a mixture of 1·0 g potassium iodide, 0·5 g iodine in 20 ml distilled water, until iodine ppt forms which persists on shaking (approx 1·5 ml is needed). Use: stain for cellulose.

Shorr's stain. 0·5 g Biebrich scarlet, 0·25 g orange G., 0·075 g fast green F.C.F., 0·5 g phosphotungstic acid, 0·5 g phosphomolybdic acid, 1 ml glacial acetic acid, 100 ml 50% ethanol. Use: vaginal smears.

Shutt's methylene blue. Mix together 100 ml ether and 100 ml pure methanol, and add 1 g methylene blue. When solution complete add 0·7 ml hydrochloric acid. Store in a stoppered bottle and keep in a cool place when not in use. Use: bacteria in milk.

Silver nitrate. 1·5% aq soln. Use: Cajal's method for Golgi bodies. For marine animals use 1% aq soln. For more detailed information see Bolles-Lee, *The Microtomist's Vade-mecum.*

Sodium metasilicate. 5% aq soln. Use: cleaning white films from glassware.

Sodium thiosulphate. Mix 0·75 g sodium thiosulphate, 10 ml 95% ethanol and 90 ml distilled water. Add a small crystal of thymol. Use: to remove iodine from iodine-treated tissue fixed in fluids containing mercuric chloride.

Solid green. See Malachite green.

Solochrome cyanine-iron alum. 0·2 g solochrome cyanine R.S., 0·5 ml 98% sulphuric acid, 95 ml distilled water, 5 ml 1% aq iron alum. Place dye in 250 ml beaker, carefully stir in acid with glass rod; effervescence occurs and a thick creamy soln is formed. Add distilled water and iron alum soln mixing thoroughly by stirring. Filter. The solution keeps well. Use: especially useful for making permanent preparations from surgical biopsies. (For method see—Hyman, J. M., Poulding, R. H., *J.M.L.T.* **18,** 107, 1961.)

Stèvenel's blue.

Two Solutions. A. 1 g methylene blue in 75 ml distilled water. B. 1·5 g potassium permanganate in 75 ml distilled water. Mix the two solutions in a flask and place in a boiling water bath until the ppt redissolves. Allow to stand for half an hour, then filter. Use: Pepine's stain for virus inclusions and elementary bodies. Also a rapid malaria stain.

Sudan black (alcoholic). Dissolve in a warm water bath 0·5 g Sudan black in 100 ml 70% ethanol. Cool and filter. Use: fat stain.

Sudan black (P.G. Soln). Dissolve by heating to 100°C for a few minutes 1 g Sudan black in 100 ml propylene glycol, cool and filter. Use: fat stain.

Sudan blue (BZL blue). Dissolve in a warm water bath 0·5 g Sudan blue in 100 ml 50% ethanol. Cool and filter. Use: fat stain.

Sudan III. Prepare as Scarlet R.

Sudan IV. See Scarlet R.

Sulphurous acid. Sat soln sulphur dioxide. Use: decalcifying fluid.

Supercedrol. Reagent for decalcifying and clearing tissues.

Susa fixative. 45 g mercuric chloride, 800 ml distilled water, 5 g sodium chloride, 20 g trichloracetic acid, 40 ml glacial acetic acid, 200 ml formalin.

Tartrazine in cellosolve. 1 g tartrazine in 100 ml cellosolve. Heat in water bath to form a sat soln. Cool and filter. Use: differentiating agent.

Terpineol. Good clearing agent for hard material.

Thioflavin S. 0·2% aq soln. Use: Negri bodies by fluorescence method.

Thionin carbol. See Carbol thionin.

Toison's blood counting fluid for white and red cells. 1 ml 1% aq gentian violet, 1 ml glacial acetic acid, add distilled water to 100 ml.

Toluidine blue. 1% aq soln. Use: Pugh and Ponder stains for bacteria (q.v.), Nissl granules, frozen sections (see also Polychrome toluidine blue). Solution consisting of 0·25 g toluidine blue in 100 ml 70% ethanol to which 0·5 ml conc hydrochloric acid has been added is used for van Wijhe's stain for cartilage.

Toluol. Clearing agent.

Tropaeolin OOO (Gold orange). 1 g in 100 ml clove oil.

Turk's blood counting fluid for white cells. 1 ml 1% aq gentian violet, 1 ml glacial acetic acid, add distilled water to 100 ml.

Twort stain. See Neutral red-light green.

Van Gieson stain (Picro fuchsin). 20 ml 1% aq acid fuchsin soln, with 1 litre sat aq picric acid soln. Use: stain for protozoa, connective tissue (Verhoeff's stain), counterstain for Ehrlich's acid haematoxylin and Weigert iron haematoxylin.

Versene. 10%. A similar chelating agent may also be used effectively (Birge and Imhoff). (See EDTA.) Use: decalcifying fluid.

Vesuvin. See Bismarck brown Y, Neisser Bismarck brown.

Water blue-orcein. 1 g water blue, 1 g orcein, 5 g glacial acetic acid, 20 ml glycerin, 50 ml ethanol, and distilled water to 100 ml. Use: connective tissue stain.

Weigert iron haematoxylin.

Two Solutions. A. 5 g haematoxylin in 500 ml ethanol. B. 12 ml strong perchloride of iron soln, 5 ml pure hydrochloric acid, add distilled water to 500 ml. Use: stain for protozoa.

Weigert's elastin stain. See Elastin stain—Weigert.

Wright blood stain. 0·3 g Wright stain in 100 ml pure methanol (prepare as Leishman stain).

Xam. Terpene resin solution—neutral, R.I. 1·5208. Use: mountant suitable for use with Romanowsky, Masson, Haematoxylin, Schiff, Thionin, Mann's stain.

Xylol. Clearing agent.

Zenker fixative. 25 g mercuric chloride, 12·5 g potassium dichromate, 5·0 g sodium sulphate, 25 ml glacial acetic acid, 500 ml distilled water. Solution keeps badly. If acetic acid is added just prior to use (omit from stock solution) this avoids deterioration.

Zenker-formol. As Zenker above but omit acetic acid and add 5 ml formalin per 100 ml just prior to use.

Zike's flagella mordant. Dissolve 20 g tannic acid in 80 ml distilled water by warming, add 50 ml sat aq soln chrome alum dropwise followed by 15 ml 1% osmic acid, 10 ml sat aq soln gentian violet.

10

Physiological Salines and Culture Solutions

Salines, culture solutions and culture media are invaluable aids for the biologist and are becoming increasingly important. They not only enable him to bathe tissues and cells for short periods whilst examination is carried out under the microscope but also to probe the physiology of whole organs, such as muscles and hearts, and to culture cells, tissues and whole small organisms so that controlled experiments and observations are possible under carefully adjusted conditions. Factors affecting growth rates, reproduction and other functions can be investigated.

A great wealth of formulae exist for media and solutions. The composition of many of them is very similar, each researcher seeming to try his own variation of a well tried recipe. Only a selection of this vast number has been listed here to cover the 'basic' ones which suffice for most general laboratory needs. Further details can be obtained in the sources listed in the bibliography at the end of the book.

1. PHYSIOLOGICAL SALINES (ANIMAL)

Clarke. 6·5 g sodium chloride, 1·4 g potassium chloride, 0·12 g calcium chloride, 0·1 g sodium bicarbonate, 0·01 g sodium hydrogen phosphate (Na_2HPO_4). Use: insects. Ordinary insect saline contains 7·5 g sodium chloride only.

Gatenby (for marine crustacea). 29·23 g sodium chloride, 0·75 g potassium chloride, 4·44 g calcium chloride.

Gatenby (for marine mollusca). 23·38 g sodium chloride, 5·55 g calcium chloride, 7·62 g magnesium chloride.

Hedon-Fleig. 7·0 g sodium chloride, 0·3 g potassium chloride, 0·1 g calcium chloride, 1·5 g sodium bicarbonate, 0·3 g magnesium sulphate. Use: invertebrates.

Holtfreter. 3·50 g sodium chloride, 0·05 g potassium chloride, 0·10 g calcium chloride, 0·20 g sodium bicarbonate. Use: amphibia.

Locke. 9·0 g sodium chloride, 0·4 g potassium chloride, 0·24 g calcium chloride, 0·2 g sodium bicarbonate in 1 litre distilled water. Use: chick embryo.

Ringer (amphibia). 6·00 g sodium chloride, 0·075 g potassium chloride, 0·10 g calcium chloride, 0·10 g sodium bicarbonate.

Ringer (chick embryo). 9·0 g sodium chloride 0·42 g potassium chloride, 0·24 g calcium chloride.

Ringer (Clarke). 6·50 g sodium chloride, 0·14 g potassium chloride, 0·12 g calcium chloride, 0·10 g sodium bicarbonate, 0·01 g sodium hydrogen phosphate (Na_2HPO_4), 2·0 g glucose. Use: amphibia. Ordinary saline contains 6·00 g sodium chloride only.

Ringer (Dale). 9·0 g sodium chloride, 0·42 g potassium chloride, 0·24 g calcium chloride, 0·15 g sodium bicarbonate. Use: mammals.

Ringer (Locke). 9·2 g sodium chloride, 0·4 g potassium chloride, 0·24 g calcium chloride, 0·5 g sodium bicarbonate (for perfusion experiments add 1·0 g glucose). Ordinary saline contains 9·0 g sodium chloride only.

Ringer (Rugh). 6·60 g sodium chloride, 0·015 g potassium chloride, 0·015 g calcium chloride. Add sufficient sodium bicarbonate to give a pH of 7·8. Use: amphibia.

Young. 13·50 g sodium chloride, 0·60 g potassium chloride, 0·25 g calcium chloride, 0·35 g magnesium chloride. Use: marine teleost.

Young. 5·50 g sodium chloride, 0·14 g potassium chloride, 0·12 g calcium chloride. Use: freshwater teleost and lamprey.

2. PLANT CULTURE SOLUTIONS

Culture solutions are used for growing a range of plants from algae to flowering plants. Often it is convenient to add 2% agar or 10% gelatine to the culture medium when growing bacteria and fungi, thus providing a 'solid' substrate for their nutrition. Flowering plants may be cultured partly immersed in the fluid or they may be fed from above or below by periodic 'watering' of the porous medium in which they are growing, as is common practice in horticultural hydroponics. There are a particularly large number of culture solutions for flowering plants the majority of which have similar constituents with only minor variations in the quantities of each. Some water-culture experiments require the absence of one different element from each culture container. Generally monovalent elements are replaced by monovalents and bivalents by bivalents, each salt being replaced with the equivalent weight of another so that the osmotic balance is not disturbed.

Artificial sea water. 29·42 g sodium chloride, 0·5 g potassium chloride, 3·22 g magnesium chloride, 0·56 g sodium bromide, 1·36 g calcium sulphate, 2·40 g magnesium sulphate, 0·11 g calcium carbonate, 0·003 g ferric oxide. Dissolve in distilled water and make up to 1 litre. This solution is equivalent in use to Naples sea water.

Osterhout's formula is as follows: 1000 parts sodium chloride, 78 parts magnesium chloride, 38 parts magnesium sulphate, 22 parts potassium chloride, 10 parts calcium chloride (gram-molecular solutions in each case). The osmotic pressure of this solution is 22·4 atmospheres.

Artificial sea water (Barnes). Dissolve the following quantities of salts in distilled water and make up to 1 litre (weight in grams). 23·991 g sodium chloride, 0·742 g potassium chloride, 1·135 g calcium chloride, 5·102 g magnesium chloride, 0·085 g sodium bromide, 0·197 g sodium bicarbonate, 4·012 g sodium sulphate, 0·011 g strontium chloride, 0·027 g boric acid. Note that the weights must be modified if hydrated salts are used.

Artificial sea water (Fowler and Allen). 27·200 g sodium chloride, 0·900 g potassium sulphate, 0·100 g calcium chloride ($CaCl_3$), 1·300 g calcium sulphate, 0·100 g magnesium bromide, 3·800 g magnesium chloride, 1·600 g magnesium sulphate. (Ref. Fowler, G. H. and Allen, E. J., *Science of the sea*, Clarendon Press, Oxford, 1928.)

Bacterial saline (Lipman). 0·8% sodium chloride, 0·04% calcium chloride, 0·04% potassium chloride, 0·08% magnesium chloride. (balanced soln of cations).

Benecke's solution. 0·5 g calcium nitrate, 0·1 g magnesium sulphate, 0·2 g potassium phosphate (KH_2PO_4) and a trace of ferric chloride. Dissolve in 1 litre of distilled water. Use: particularly for algae.

Bristol's solution. 1 g potassium dihydrogen phosphate, 1 g sodium nitrate, 0·3 g magnesium sulphate, 0·1 g calcium chloride 0·1 g sodium chloride, trace ferric chloride in 1 litre distilled water. Use: Culture of algae.

Brown's medium. 2·0 g agar, 0·20 g asparagen, 0·075 g magnesium sulphate, 0·125 g potassium phosphate, 0·20 g glucose, 1·00 g starch in 100 ml distilled water. Use: Culture of fungi.

Cohn's solution. 5·0 g magnesium sulphate, 5·0 g potassium phosphate (KH_2PO_4), 10·0 g ammonium tartrate, 0·5 g potassium chloride, dissolved in 1 litre of distilled water.

Dung agar. Soak 1000 g horse or cow dung for 3 days in cold water. Decant and add distilled water until mixture is straw coloured. Add 2·5 g agar for each 100 ml of this extract. Use: culture of fungi.

Hansen's medium. 1 g peptone, 5·9 g maltose, 0·3 g potassium phosphate (KH_2PO_4), 0·2 g magnesium sulphate in 100 ml distilled water. Use: medium for culture of yeasts. 5 g dextrose may be substituted for maltose and 0·5 g magnesium sulphate for 0·2 g in an alternative formula.

Hay infusion. Place 10 g chopped hay in 1 litre distilled water. Heat to 70°C for 45 min. Filter. To make soln neutral to litmus add a few drops of sodium hydroxide. Cool. Use: culture of protozoa.

Hoagland and Snyder. For trace and other accessory elements in water culture solutions. Dissolve the following in 18 litres of water. 1·0 g aluminium sulphate, 1·0 g cobalt nitrate, 1·0 g copper sulphate, 11·0 g boric acid, 0·5 g lithium chloride, 7·0 g manganese chloride, 1·0 g nickel sulphate, 0·5 g potassium bromide, 0·5 g potassium iodide, 0·5 g stannous chloride, 1·0 g titanium oxide, 1·0 g zinc sulphate. To each litre of normal culture solution add 1 ml of this solution.

Knop's solution. 5 g potassium chloride, 1 g potassium nitrate, 1 g potassium phosphate (KH_2PO_4), 4 g calcium nitrate ($Ca(NO_3)_2.4H_2O$), 1 g magnesium sulphate ($MgSO_4.7H_2O$), trace ferric chloride. Dissolve in 12 litres distilled water. Use: flowering plants.

For algae make up two solutions. A. Dissolve in 1 litre of

distilled water 1 g magnesium sulphate, 1 g potassium nitrate, 1 g potassium phosphate (K_2HPO_2). B. 4 g calcium nitrate in 1 litre of distilled water. Add B to A. Solution useful for producing fruiting bodies of green algae.

Litmus milk. Add 2 ml sat litmus soln to 100 ml skim milk. Sterilize in autoclave for 15 min. Use: bacterial culture.

Mayer's fluid. Dissolve in 1 litre distilled water, 10·0 g potassium phosphate (KH_2PO_4), 10·0 g magnesium sulphate, 15·0 g ammonium nitrate, 0·1 g calcium phosphate ($Ca^3(PO_4)_2$). Add suitable sugar to provide carbon.

Pasteur (modified). Dissolve in 2 litres of distilled water, 150 g glucose, 2·0 g potassium phosphate (KH_2PO_4), 10·0 g ammonium tartrate, 0·2 g calcium phosphate ($CaHPO_4$), 2·0 g potassium phosphate (KH_2PO_4). Use: cultivation of yeast.

Pfeffer's solution. Dissolve in 3–7 litres of distilled water, 4 g calcium nitrate, 1 g potassium nitrate, 1 g magnesium sulphate, 1 g potassium phosphate (KH_2PO_4), 5 g potassium chloride, trace ferric chloride. Use: flowering plant culture. If solutions deficient in one element are required replace calcium by potassium, potassium by calcium, magnesium by potassium, nitrate by chloride, sulphate by chloride, phosphate by sulphate: for iron free soln omit ferric chloride.

Potato agar. Chop up 200 g peeled potatoes into 1 litre distilled water and boil for 15 mins. Strain off liquid, make up to 1 litre with distilled water, add 15 g agar. Boil to dissolve, filter. Use: culture of bacteria.

Richard's solution. Dissolve in 1 litre of distilled water, 6·6 g potassium nitrate, 3·3 g potassium dihydrogen phosphate, 33·3 g sucrose, 1·7 g magnesium sulphate. Use: culture of moulds.

Ringer. Dissolve in 1 litre of distilled water, 7·5 g sodium chloride, 0·075 g potassium chloride, 0·1 g calcium chloride, 0·1 g sodium bicarbonate. Use: observation of living cells.

Sachs' culture solution. Dissolve in 1 litre of distilled water, 0·25 g calcium sulphate ($CaSO_4.2H_2O$), 0·25 g calcium phosphate ($CaH_4(PO_4)_2.2H_2O$), 0·25 g magnesium sulphate ($MgSO_4.7H_2O$), 0·08 g sodium chloride, 0·70 g potassium nitrate, 0·005 g iron chloride ($FeCl^3.6H_2O$). For solutions deficient in one element use the following substitutions: replace calcium sulphate by 0·20 g potassium sulphate, and calcium phosphate by sodium phosphate 0·71 g: omit ferric chloride: replace potassium nitrate by 0·52 g potassium chloride: replace calcium phosphate by 0·16 g calcium nitrate: replace calcium sulphate by 0·16 g calcium chloride and magnesium sulphate by 0·21 g magnesium chloride: replace magnesium sulphate by 0·17 g potassium sulphate: replace potassium nitrate by 0·59 g sodium nitrate.

II

Miscellaneous Solutions

Soap solution for surface tension (Searle's). Dissolve 18·75 g of sodium oleate in 750 ml of cold distilled water by vigorous shaking. Allow to settle for a week in a dark cupboard and siphon off the clear liquid. Keep in a well-stoppered bottle in the dark.

Soap solution for bubbles. Shake 4 g of Castile soap with 80 ml of water and add 20 ml of glycerine.

Solution for simple manometers. Add 10 drops of fluorescein soln 1 g/100 ml ethanol (see indicators) to 1 litre of water. This solution or a slightly stronger one can be used for showing the path of a beam of light in refraction experiments.

Solution to silver glass. Add 2 M ammonia soln to 100 ml of freshly prepared silver nitrate soln (0·1 M) until the ppt first formed just redissolves. Add 7·5 ml of 2 M sodium hydroxide soln and redissolve the ppt with ammonia soln and produce a faint reprecipitation of silver oxide by adding a few drops of the 0·1 M silver nitrate soln. The addition of a small quantity of glucose is recommended by some formulae.

Water culture solution. Dissolve in 1 litre of water 1 g of potassium nitrate, 0·5 g of potassium dihydrogen phosphate, 0·5 g calcium sulphate, 0·5 g magnesium sulphate, 0·5 g of sodium chloride and add two drops of ferric chloride soln.

SOLUTIONS FOR MAKING INDICATOR PAPERS

Starch-potassium iodide. Make 1 g of soluble starch into a smooth thin paste with a little water, pour it into 100 ml of boiling water and dissolve 1 g of potassium iodide in the soln.

Heat-sensitive paper. Dissolve 1·5 g of cobalt chloride crystals and 1·5 g of calcium chloride in 1 litre of water.

Potassium cobalticyanide. Test for zinc. Dissolve 4 g of potassium cobalticyanide and 1 g of potassium chlorate in 100 ml of water.

Ferrox reagent. Test for iron. Shake 1 g of ferric chloride and 1 g of potassium thiocyanate separately in 10 ml portions of methanol, mix the solns and filter. Dip strips of paper twice in the soln, drying after each dipping.

Turmeric. Digest ground turmeric root with water several times, discarding the soln in each case. Dry the residue and shake it for some days at intervals in a stoppered bottle with its own weight of ethanol and filter the soln.

In each case the soln is used to soak either filter paper or duplicating paper which is then dried.

ELECTROLYTE SOLUTIONS FOR CELLS AND ELECTROLYSIS

Accumulator acid (sulphuric acid). Pour carefully, with stirring, 220 ml of conc sulphuric acid into approx 750 ml of distilled water and make up to 1 litre. Alternatively, add 303 ml of conc acid in the same way to 1 litre of water. Check the specific gravity of the acid with a hydrometer and adjust it, if necessary, to 1·25. (More acid increases and more water decreases the specific gravity.)

Daniell cell. (*a*) 80 ml of conc sulphuric acid added to water and made up to 1 litre (*b*) Sat copper sulphate soln containing approx 400 g of crystals per litre. Add approx 2 ml of conc sulphuric acid.

Leclanché cell. A nearly sat soln of ammonium chloride containing 350 g per litre.

Copper sulphate solution. For electrolysis. Dissolve 150 g of copper sulphate crystals in 1 litre of water and add 25 ml of conc sulphuric acid and 60 ml of ethanol.

Sulphuric acid. For electrolysis. Add 3 ml of conc acid for every 100 ml of soln needed.

Silver nitrate solution. For electrolysis. Use M soln containing approx 170 g per litre.

Solution to platinize platinum electrodes. Pour 10 ml of 5% platinum chloride soln and 1 ml of 0·1 M lead acetate soln into 25 ml of 4 M hydrochloric acid and make up to 50 ml with water. Use this as the electrolyte and the platinum electrodes reversing the current occasionally for about 20 minutes.

Appendix

MAXIMUM TOLERANCES *or errors in Graduated Glassware*:

Volume		5 ml	10 ml	20 ml	25 ml	50 ml	100 ml	250 ml
Burette	A	0·02	0·01		0·03	0·05		
	B	0·04	0·02		0·06	0·10		
Pipette	A	0·015	0·02	0·03	0·03	0·04		
	B	0·03	0·04	0·06	0·06	0·08		

Volume		25 ml	50 ml	100 ml	250 ml	500 ml	1000 ml
Flask	A	0·03	0·05	0·08	0·15	0·25	0·40
	B	0·06	0·10	0·15	0·30	0·50	0·80

	0	1	2	3	4	5	6	7	8	9	1 2 3	4 5 6	7 8 9
10	0000	0043	0086	0128	0170						5 9 13	17 21 26	30 34 38
						0212	0253	0294	0334	0374	4 8 12	16 20 24	28 32 36
11	0414	0453	0492	0531	0569						4 8 12	16 20 23	27 31 35
						0607	0645	0682	0719	0755	4 7 11	15 18 22	26 29 33
12	0792	0828	0864	0899	0934						3 7 11	14 18 21	25 28 32
						0969	1004	1038	1072	1106	3 7 10	14 17 20	24 27 31
13	1139	1173	1206	1239	1271						3 6 10	13 16 19	23 26 29
						1303	1335	1367	1399	1430	3 7 10	13 16 19	22 25 29
14	1461	1492	1523	1553	1584						3 6 9	12 15 19	22 25 28
						1614	1644	1673	1703	1732	3 6 9	12 14 17	20 23 26
15	1761	1790	1818	1847	1875						3 6 9	11 14 17	20 23 26
						1903	1931	1959	1987	2014	3 6 8	11 14 17	19 22 25
16	2041	2068	2095	2122	2148						3 6 8	11 14 16	19 22 24
						2175	2201	2227	2253	2279	3 5 8	10 13 16	18 21 23
17	2304	2330	2355	2380	2405						3 5 8	10 13 15	18 20 23
						2430	2455	2480	2504	2529	3 5 8	10 12 15	17 20 22
18	2553	2577	2601	2625	2648						2 5 7	9 12 14	17 19 21
						2672	2695	2718	2742	2765	2 4 7	9 11 14	16 18 21
19	2788	2810	2833	2856	2878						2 4 7	9 11 13	16 18 20
						2900	2923	2945	2967	2989	2 4 6	8 11 13	15 17 19
20	3010	3032	3054	3075	3096	3118	3139	3160	3181	3201	2 4 6	8 11 13	15 17 19
21	3222	3243	3263	3284	3304	3324	3345	3365	3385	3404	2 4 6	8 10 12	14 16 18
22	3424	3444	3464	3483	3502	3522	3541	3560	3579	3598	2 4 6	8 10 12	14 15 17
23	3617	3636	3655	3674	3692	3711	3729	3747	3766	3784	2 4 6	7 9 11	13 15 17
24	3802	3820	3838	3856	3874	3892	3909	3927	3945	3962	2 4 5	7 9 11	12 14 16
25	3979	3997	4014	4031	4048	4065	4082	4099	4116	4133	2 3 5	7 9 10	12 14 15
26	4150	4166	4183	4200	4216	4232	4249	4265	4281	4298	2 3 5	7 8 10	11 13 15
27	4314	4330	4346	4362	4378	4393	4409	4425	4440	4456	2 3 5	6 8 9	11 13 14
28	4472	4487	4502	4518	4533	4548	4564	4579	4594	4609	2 3 5	6 8 9	11 12 14
29	4624	4639	4654	4669	4683	4698	4713	4728	4742	4757	1 3 4	6 7 9	10 12 13
30	4771	4786	4800	4814	4829	4843	4857	4871	4886	4900	1 3 4	6 7 9	10 11 13
31	4914	4928	4942	4955	4969	4983	4997	5011	5024	5038	1 3 4	6 7 8	10 11 12
32	5051	5065	5079	5092	5105	5119	5132	5145	5159	5172	1 3 4	5 7 8	9 11 12
33	5185	5198	5211	5224	5237	5250	5263	5276	5289	5302	1 3 4	5 6 8	9 10 12
34	5315	5328	5340	5353	5366	5378	5391	5403	5416	5428	1 3 4	5 6 8	9 10 11
35	5441	5453	5465	5478	5490	5502	5514	5527	5539	5551	1 2 4	5 6 7	9 10 11
36	5563	5575	5587	5599	5611	5623	5635	5647	5658	5670	1 2 4	5 6 7	8 10 11
37	5682	5694	5705	5717	5729	5740	5752	5763	5775	5786	1 2 3	5 6 7	8 9 10
38	5798	5809	5821	5832	5843	5855	5866	5877	5888	5899	1 2 3	5 6 7	8 9 10
39	5911	5922	5933	5944	5955	5966	5977	5988	5999	6010	1 2 3	4 5 7	8 9 10
40	6021	6031	6042	6053	6064	6075	6085	6096	6107	6117	1 2 3	4 5 6	8 9 10
41	6128	6138	6149	6160	6170	6180	6191	6201	6212	6222	1 2 3	4 5 6	7 8 9
42	6232	6243	6253	6263	6274	6284	6294	6304	6314	6325	1 2 3	4 5 6	7 8 9
43	6335	6345	6355	6365	6375	6385	6395	6405	6415	6425	1 2 3	4 5 6	7 8 9
44	6435	6444	6454	6464	6474	6484	6493	6503	6513	6522	1 2 3	4 5 6	7 8 9
45	6532	6542	6551	6561	6571	6580	6590	6599	6609	6618	1 2 3	4 5 6	7 8 9
46	6628	6637	6646	6656	6665	6675	6684	6693	6702	6712	1 2 3	4 5 6	7 7 8
47	6721	6730	6739	6749	6758	6767	6776	6785	6794	6803	1 2 3	4 5 5	6 7 8
48	6812	6821	6830	6839	6848	6857	6866	6875	6884	6893	1 2 3	4 4 5	6 7 8
49	6902	6911	6920	6928	6937	6946	6955	6964	6972	6981	1 2 3	4 4 5	6 7 8

	0	1	2	3	4	5	6	7	8	9	1	2	3	4	5	6	7	8	9
50	6990	6998	7007	7016	7024	7033	7042	7050	7059	7067	1	2	3	3	4	5	6	7	8
51	7076	7084	7093	7101	7110	7118	7126	7135	7143	7152	1	2	3	3	4	5	6	7	8
52	7160	7168	7177	7185	7193	7202	7210	7218	7226	7235	1	2	2	3	4	5	6	7	7
53	7243	7251	7259	7267	7275	7284	7292	7300	7308	7316	1	2	2	3	4	5	6	6	7
54	7324	7332	7340	7348	7356	7364	7372	7380	7388	7396	1	2	2	3	4	5	6	6	7
55	7404	7412	7419	7427	7435	7443	7451	7459	7466	7474	1	2	2	3	4	5	5	6	7
56	7482	7490	7497	7505	7513	7520	7528	7536	7543	7551	1	2	2	3	4	5	5	6	7
57	7559	7566	7574	7582	7589	7597	7604	7612	7619	7627	1	2	2	3	4	5	5	6	7
58	7634	7642	7649	7657	7664	7672	7679	7686	7694	7701	1	1	2	3	4	4	5	6	7
59	7709	7716	7723	7731	7738	7745	7752	7760	7767	7774	1	1	2	3	4	4	5	6	7
60	7782	7789	7796	7803	7810	7818	7825	7832	7839	7846	1	1	2	3	4	4	5	6	6
61	7853	7860	7868	7875	7882	7889	7896	7903	7910	7917	1	1	2	3	4	4	5	6	6
62	7924	7931	7938	7945	7952	7959	7966	7973	7980	7987	1	1	2	3	3	4	5	6	6
63	7993	8000	8007	8014	8021	8028	8035	8041	8048	8055	1	1	2	3	3	4	5	5	6
64	8062	8069	8075	8082	8089	8096	8102	8109	8116	8122	1	1	2	3	3	4	5	5	6
65	8129	8136	8142	8149	8156	8162	8169	8176	8182	8189	1	1	2	3	3	4	5	5	6
66	8195	8202	8209	8215	8222	8228	8235	8241	8248	8254	1	1	2	3	3	4	5	5	6
67	8261	8267	8274	8280	8287	8293	8299	8306	8312	8319	1	1	2	3	3	4	5	5	6
68	8325	8331	8338	8344	8351	8357	8363	8370	8376	8382	1	1	2	3	3	4	4	5	6
69	8388	8395	8401	8407	8414	8420	8426	8432	8439	8445	1	1	2	2	3	4	4	5	6
70	8451	8457	8463	8470	8476	8482	8488	8494	8500	8506	1	1	2	2	3	4	4	5	6
71	8513	8519	8525	8531	8537	8543	8549	8555	8561	8567	1	1	2	2	3	4	4	5	5
72	8573	8579	8585	8591	8597	8603	8609	8615	8621	8627	1	1	2	2	3	4	4	5	5
73	8633	8639	8645	8651	8657	8663	8669	8675	8681	8686	1	1	2	2	3	4	4	5	5
74	8692	8698	8704	8710	8716	8722	8727	8733	8739	8745	1	1	2	2	3	4	4	5	5
75	8751	8756	8762	8768	8774	8779	8785	8791	8797	8802	1	1	2	2	3	3	4	5	5
76	8808	8814	8820	8825	8831	8837	8842	8848	8854	8859	1	1	2	2	3	3	4	5	5
77	8865	8871	8876	8882	8887	8893	8899	8904	8910	8915	1	1	2	2	3	3	4	4	5
78	8921	8927	8932	8938	8943	8949	8954	8960	8965	8971	1	1	2	2	3	3	4	4	5
79	8976	8982	8987	8993	8998	9004	9009	9015	9020	9025	1	1	2	2	3	3	4	4	5
80	9031	9036	9042	9047	9053	9058	9063	9069	9074	9079	1	1	2	2	3	3	4	4	5
81	9085	9090	9096	9101	9106	9112	9117	9122	9128	9133	1	1	2	2	3	3	4	4	5
82	9138	9143	9149	9154	9159	9165	9170	9175	9180	9186	1	1	2	2	3	3	4	4	5
83	9191	9196	9201	9206	9212	9217	9222	9227	9232	9238	1	1	2	2	3	3	4	4	5
84	9243	9248	9253	9258	9263	9269	9274	9279	9284	9289	1	1	2	2	3	3	4	4	5
85	9294	9299	9304	9309	9315	9320	9325	9330	9335	9340	1	1	2	2	3	3	4	4	5
86	9345	9350	9355	9360	9365	9370	9375	9380	9385	9390	1	1	2	2	3	3	4	4	5
87	9395	9400	9405	9410	9415	9420	9425	9430	9435	9440	0	1	1	2	2	3	3	4	4
88	9445	9450	9455	9460	9465	9469	9474	9479	9484	9489	0	1	1	2	2	3	3	4	4
89	9494	9499	9504	9509	9513	9518	9523	9528	9533	9538	0	1	1	2	2	3	3	4	4
90	9542	9547	9552	9557	9562	9566	9571	9576	9581	9586	0	1	1	2	2	3	3	4	4
91	9590	9595	9600	9605	9609	9614	9619	9624	9628	9633	0	1	1	2	2	3	3	4	4
92	9638	9643	9647	9652	9657	9661	9666	9671	9675	9680	0	1	1	2	2	3	3	4	4
93	9685	9689	9694	9699	9703	9708	9713	9717	9722	9727	0	1	1	2	2	3	3	4	4
94	9731	9736	9741	9745	9750	9754	9759	9763	9768	9773	0	1	1	2	2	3	3	4	4
95	9777	9782	9786	9791	9795	9800	9805	9809	9814	9818	0	1	1	2	2	3	3	4	4
96	9823	9827	9832	9836	9841	9845	9850	9854	9859	9863	0	1	1	2	2	3	3	4	4
97	9868	9872	9877	9881	9886	9890	9894	9899	9903	9908	0	1	1	2	2	3	3	4	4
98	9912	9917	9921	9926	9930	9934	9939	9943	9948	9952	0	1	1	2	2	3	3	4	4
99	9956	9961	9965	9969	9974	9978	9983	9987	9991	9996	0	1	1	2	2	3	3	3	4

ATOMIC WEIGHTS (*carbon scale*)

	Symbol.	At. wt.*		Symbol.	At. wt.*
Actinum	Ac	227	Molybdenum	Mo	95·94
Aluminium	Al	26·9815	Neodymium	Nd	144·24
Americium	Am	[243]	Neon	Ne	20·183
Antimony	Sb	121·75	Neptunium	Np	[237]
Argon	Ar	39·948	Nickel	Ni	58·71
Arsenic	As	74·9216	Niobium	Nb	92·906
Astatine	At	[210]	Nitrogen	N	14·0067
Barium	Ba	137·34	Osmium	Os	190·2
Berkelium	Bk	[249]	Oxygen	O	15·9994
Beryllium	Be	9·0122	Palladium	Pd	106·4
Bismuth	Bi	208·980	Phosphorus	P	30·9738
Boron	B	10·811	Platinum	Pt	195·09
Bromine	Br	79·909	Plutonium	Pu	[242]
Cadmium	Cd	112·40	Polonium	Po	210
Caesium	Cs	132·905	Potassium	K	39·102
Calcium	Ca	40·08	Praseodymium	Pr	140·907
Californium	Cf	[249]	Promethium	Pm	[145]
Carbon	C	12·0115	Protactinium	Pa	231
Cerium	Ce	140·12	Radium	Ra	226·05
Chlorine	Cl	35·543	Radon	Rn	222
Chromium	Cr	51·996	Rhenium	Re	186·2
Cobalt	Co	58·9332	Rhodium	Rh	102·905
Copper	Cu	63·54	Rubidium	Rb	85·47
Curium	Cm	[245]	Ruthenium	Ru	101·07
Dysprosium	Dy	162·50	Samarium	Sm	150·35
Erbium	Er	167·26	Scandium	Sc	44·965
Europium	Eu	151·96	Selenium	Se	78·96
Fluorine	F	18·9984	Silicon	Si	28·086
Francium	Fr	[223]	Silver	Ag	107·870
Gadolinium	Gd	157·45	Sodium	Na	22·9898
Gallium	Ga	69·72	Strontium	Sr	87·62
Germanium	Ge	72·59	Sulphur	S	32·064
Gold	Au	196·967	Tantalum	Ta	180·948
Hafnium	Hf	178·49	Technetium	Tc	[99]
Helium	He	4·0026	Tellurium	Te	127·60
Holmium	Ho	164·930	Terbium	Tb	158·924
Hydrogen	H	1·00797	Thallium	Tl	204·37
Indium	In	114·82	Thorium	Th	232·038
Iodine	I	126·9044	Thulium	Tm	168·934
Iridium	Ir	192·2	Tin	Sn	118·69
Iron	Fe	55·847	Titanium	Ti	47·90
Krypton	Kr	83·80	Tungsten	W	183·85
Lanthanum	La	138·91	Uranium	U	238·03
Lead	Pb	207·19	Vanadium	V	50·942
Lithium	Li	6·939	Xenon	Xe	111·30
Lutetium	Lu	174·97	Ytterbium	Yb	173·04
Magnesium	Mg	24·312	Yttrium	Y	88·905
Manganese	Mn	54·9380	Zinc	Zn	65·37
Mercury	Hg	200·59	Zirconium	Zr	91·22

* A value in brackets [ ] is mass number of the most stable isotope known.

Simple First Aid Procedures

The following details apply to the treatment of injuries most likely to be sustained in a laboratory where chemicals are handled. They cover not only the recommended procedures for injuries due to chemicals but also burns, cuts and the like. It must be emphasized that the treatments suggested are essentially *immediate* ones. They should not be regarded as substitutes for professional attention from a doctor or a trained nurse who should be called immediately if necessary. With extremely toxic chemicals such as cyanides speed is critical.

FIRST AID BOX

It is recommended that a laboratory should maintain a first aid box containing the following (in addition to providing sterilized dressings, bandages, etc.), a 16 oz eye irrigation bottle, tablespoon, jar of common salt, jar of magnesium sulphate (Epsom salts), milk of magnesia, bottle of vinegar or 1% acetic acid.

The following remedies and antidotes are recommended for specific chemicals:

Bromine, formic acid and hydrofluoric acid. Splashed on skin. A 12 oz bottle of dilute ammonia solution (1 vol 0·88 sg ammonia in 15 vols of water) should be kept. Apply soln to affected area.

Cyanides. Give cyanide antidote, which is made up as follows. The antidote consists of two solutions. Antidote A—dissolve 158 g pure ferrous sulphate crystals ($FeSO_4.7H_2O$) and 3 g pure citric acid crystals in 1 litre distilled water (inspect regularly and change if necessary). Antidote B—dissolve 60 g anhydrous sodium carbonate in 1 litre distilled water. Place 50 ml of A and B in separate wide-mouth bottles and stopper in polythene covered stoppers. Label clearly *mix A and B before use and swallow the mixture.* If the affected person is breathing, break an amyl nitrite capsule and give to inhale for 15–30 sec, repeating every 2–3 min. If breathing has stopped apply artificial respiration. *Obtain medical attention.*

Emetics. Useful emetics are salt water (one tablespoonful common salt in a tumblerful of warm water) and a thick cream of mustard powder in warm water.

Hydrogen cyanide. Inhaled by gassing casualties. Amyl nitrite capsules (3 minims).

Iodine. After ingestion or skin contact. Keep sodium thiosulphate crystals in a bottle ready to prepare a fresh 1% soln.

Phosphorus. Skin burns. A 12 oz bottle of 3% copper sulphate soln.

The following additional notes may also be useful

Strong alkalis. If sucked up into the mouth, wash out thoroughly with water, give plenty of water to drink followed by vinegar or 1% acetic acid. If the eyes or skin are affected also wash thoroughly with water and bathe the eyes with 4% boric acid.

Strong acids. If sucked up into the mouth, wash out thoroughly with water or 5% sodium carbonate. Give plenty of water to drink followed by milk of magnesia if the acid is actually swallowed (this is most unlikely). The eyes and skin should also be thoroughly washed in water followed by 5% sodium bicarbonate for the eyes and 5% ammonia for the skin.

Burns and scalds (from heat). If serious apply a dry sterilized dressing. Cover extensive areas with a clean towel loosely applied. Do not pierce blisters or remove dressings that stick to the affected area.

The information contained in these brief notes is largely based on the laboratory first aid chart supplied by the British Drug Houses Limited which in turn was largely prepared from *The Laboratory Handbook of Toxic Agents* published by the Royal Institute of Chemistry.

Bibliography

Quantitative Inorganic Analysis—Belcher and Nutten. Butterworth

Text book of Quantitative Inorganic Analysis—Kolthoff and Sandell. Macmillan

Textbook of Quantitative Inorganic Analysis—Vogel. Longman

Quantitative Chemical Analysis—Clowes and Coleman (ed. Grant). Churchill

Comprehensive Analytical Chemistry—Wilson and Wilson. Elsevier

Fenton's Notes on Qualitative Analysis—Saunders. C.U.P.

Systematic Organic Analysis—Middleton. Arnold

Handbook of Chemistry—Lange. Handbook Publishers Inc.

A Reference Book of Chemistry—White. U.L.P.

School Laboratory Management—Sutcliffe. Arnold

Analytical Applications of Diamino-ethane-tetra-acetic acid. British Drug Houses Ltd

The Laboratory Chemicals Catalogues of British Drug Houses Ltd and of May & Baker Ltd.

McClung's Handbook of Microscopical Technique—Ruth McClung Jones. Hafner

Notes on Microscopical techniques for Zoologists—C. F. A. Pantin. C.U.P.

Elementary Microtechnique—H. Alan Peacock. Arnold

Plant Science Formulae—McLean and Cook. Macmillan

Biological Staining Methods—George T. Gurr. Gurr

The Microtomist's Vade-mecum—Bolles Lee. ed. Gatenby and Beams

Laboratory Techniques in Zoology—Mahoney. Butterworth

Biological Stains, 7th edition—H. J. Conn. Williams & Wilkins

Botanical Histochemistry—William A. Jensen. Freeman

Microbiological Methods—Collins. Butterworth

Principles of Biological Microtechnique—J. R. Baker. Methuen

Cytological Technique, 4th edition—J. R. Baker. Methuen

Encyclopaedia of Microscopic Stains—E. Gurr. Leonard Hill

Staining Animal Tissues—E. Gurr. Leonard Hill
Handbook of Histopathological Technique—C. F. A. Calling. Butterworth
Histological and Histochemical Techniques—H. A. Davenport. W. B. Saunders
Animal Tissue Techniques—Humason. Freeman
Handbook of Medical Laboratory Formulae—R. E. Silverton and M. J. Anderson. Butterworth.
Histological Technique—H. M. Carleton and R. A. B. Drury.
Histochemistry—A. G. E. Pearse. Little.
Staining Procedures—Conn, Darrow and Emmel.Williams & Wilkins.

Index